ROAD TRANSPORT RESEARCH

congestion control and demand management

REPORT PREPARED BY
AN OECD SCIENTIFIC EXPERT GROUP

ORGANISATION FOR ECONOMIC CO-OPERATION AND DEVELOPMENT

ORGANISATION FOR ECONOMIC CO-OPERATION AND DEVELOPMENT

Pursuant to Article 1 of the Convention signed in Paris on 14th December 1960, and which came into force on 30th September 1961, the Organisation for Economic Co-operation and Development (OECD) shall promote policies designed:

— to achieve the highest sustainable economic growth and employment and a rising standard of living in Member countries, while maintaining financial stability, and thus to contribute to the development of the world economy;

— to contribute to sound economic expansion in Member as well as non-member countries in the process of economic development; and

— to contribute to the expansion of world trade on a multilateral, non-discriminatory basis in accordance with international obligations.

The original Member countries of the OECD are Austria, Belgium, Canada, Denmark, France, Germany, Greece, Iceland, Ireland, Italy, Luxembourg, the Netherlands, Norway, Portugal, Spain, Sweden, Switzerland, Turkey, the United Kingdom and the United States. The following countries became Members subsequently through accession at the dates indicated hereafter: Japan (28th April 1964), Finland (28th January 1969), Australia (7th June 1971), New Zealand (29th May 1973) and Mexico (18th May 1994). The Commission of the European Communities takes part in the work of the OECD (Article 13 of the OECD Convention).

Publié en français sous le titre :

GÉRER LA CONGESTION ET LA DEMANDE DE TRAFIC ROUTIER

Photographer: Courtesy P. J. Santini, CETUR

FOREWORD

The Programme centres on road and road transport research, while taking into account the impacts of intermodal aspects on the road transport system as a whole. It is geared towards a technico-economic approach to solving key road transport issues identified by Member countries. The Programme has two main fields of activity:

-- International research and policy assessments of road and road transport issues to provide scientific support for decisions by Member governments and international governmental organisations;

-- Technology transfer and information exchange through two databases - the International Road Research Documentation (IRRD) scheme and the International Road Traffic and Accident Database (IRTAD).

The scientific and technical activities concern:

-- The management, rehabilitation and environmental assessment of road and bridge infrastructure;

-- The formulation and evaluation of targeted road and traffic safety programmes;

-- The development and management of road traffic technology and advanced driver communication systems;

-- The assessment of urban and inter-urban road transport strategies, freight operations and logistics approaches;

-- The strategic planning and management of research and joint projects as well as technology diffusion, both in OECD countries and economies in transition;

-- The maintenance management of road infrastructure and the evaluation of traffic safety measures and strategies in developing countries.

ABSTRACT

IRRD No. 865266

This study was carried out by an OECD Scientific Expert Group. Its aim was to share information between Member countries on road congestion counter measures. It reviews policies as well as conventional and innovative measures applied to solve congestion problems. Chapter I describes the current background of congestion problems which are becoming more and more severe. The necessity to apply demand or supply measures is highlighted. The report identifies the factors on which these measures can act. Chapter II presents a detailed catalogue of congestion management measures presented in nine sections, each representing a particular strategy class : Land Use and Zoning; Telecommunications Substitutes; Traveller Information Services; Economic Measures; Administrative Measures; Road Traffic Operations; Preferential Treatment; Public Transport Operations; Freight Movements.   Chapter III reviews existing policies, plans and programmes for an effective implementation of congestion management measures. It is illustrated by several concrete case studies. Chapter IV deals with the future of congestion management and with the influence of new policies and technologies. Finally Chapter V presents the conclusions and recommendations of the Expert Group. This report offers a large source of information for policy makers and administrators responsible for congestion management and will assist countries and cities in their effort of implementing relevant measures to solve their specific road congestion problems.

Field Classification: Traffic and Transport; Traffic control; traffic and transport planning.

Field Codes: 70; 72; 73

Keywords: Car pooling; congestion (traffic); journey to work; traffic control ; traffic regulation ; peak hour; land use; planning; policy; public transport; operations (public transport); road pricing; transport.

TABLE OF CONTENTS

EXECUTIVE SUMMARY

AIMS OF THE STUDY

This Report was prepared by the OECD Expert Group on "*Congestion Control and Demand Management*" to share information amongst Member countries on the broad collection of measures being applied to reduce the impact of road traffic congestion problems. The Report is designed to show what conventional and innovative measures are being taken to address the increasingly difficult traffic congestion issues.

THE CHALLENGE

Road traffic congestion is a significant problem in many parts of the world. Congestion continues to increase and the conventional approach of "building more roads" will not always be a solution to the problem for political, financial, and environmental reasons. In addition, it has been found that building new roads can, in some cases, compound the congestion problem by inducing greater unforeseen demands for travel by auto that quickly fills the additional capacity.

Road traffic congestion is not simply a problem confined to commuter trips in large cities or urban areas. Congestion affects the work trip and the non-work trip. It affects the movement of people and the flow of goods. In non-urban (rural) areas and inter-city corridors, traffic is disrupted by incidents, maintenance operations, detours, over-loaded tourist routes, and other causes. From the road traveller's point-of-view, traffic delays are substantial and growing. From the employers point-of-view, congestion takes its toll in lost worker productivity, delivery delays, and cost. Speed, reliability, and cost of urban and intercity freight movements are increasingly affected by congestion.

The causes of congestion can be categorized as either recurring or non-recurring. Recurring congestion is the predictable delay caused by high volumes of vehicles using the roadway during the same daily time periods (i.e. peak commute periods, holiday periods, or special events) and at peak locations (i.e. intersections, interchanges, major long term construction areas, or toll plaza areas). Non-recurring congestion is unpredictable delay, generally caused by spontaneous, unplanned occurrences, e.g. traffic accidents and incidents, emergency maintenance, or weather conditions.

Given the growing body of evidence to show that traffic congestion problems cannot be solved simply by expanding the road infrastructure alone, many countries are trying to manage their existing transport systems in order to enhance mobility and safety, reduce demand for car use, and improve traffic fluidity. Experience and technical research have demonstrated that when properly applied, measures taken to manage the existing transport system can have a profound impact on trip making behaviour and traffic congestion. A variety of measures and technologies for motorised and non-motorised transport as well as economic and administrative policies have been used to manage congestion and influence travel demand. Many transport professionals also believe that advancements in new technologies will be applied to help manage congestion in the future.

DEMAND-SIDE AND SUPPLY-SIDE STRATEGIES

The collection of measures are termed "Congestion Management" and presented in two strategy classifications: demand-side and supply-side. Demand-side congestion management measures are designed to reduce car dependency and car demand on the system by increasing vehicle occupancy, increasing public transport mode share, reducing the need to travel during a specified peak time period, and/or reducing the need to travel to a specified location. Supply-side congestion management measures are intended to increase the existing capacity of the system in order to improve traffic flow for all modes.

CONGESTION MANAGEMENT OBJECTIVES

Congestion management measures are designed to improve the operating efficiency of the existing transport system (infrastructure, modes, and services) in three ways. First, congestion management offers the opportunity to increase the use of alternative transport modes including public transport, carpooling, and bicycle/walking. Second, congestion management offers the opportunity to alter trip patterns through the application of measures like land use policies, alternative work schedule arrangements, telecommuting, and pricing. Third, congestion management offers the opportunity to improve traffic flow through measures like traffic signal improvements, incident management and route guidance system.

Implemented individually or in concert with one another, congestion management measures can help to achieve one or more of eight clearly measurable objectives. These objectives, which can also be considered as positive impacts, are listed as follows:

1. reducing the need to make a trip,
2. reducing the length of a trip,
3. promoting non-motorised transport,
4. promoting public transport,
5. promoting carpooling,
6. shifting peak-hour travel,
7. shifting travel from congested locations, and
8. reducing traffic delays.

Congestion management measures can be considered throughout a metropolitan area to address system, corridor, individual facility, and site improvements. They can also be implemented along major intercity corridors to relieve traffic congestion problems caused by work travel, construction, holiday or vacation travel, freight movement, and weather.

CONTENTS OF REPORT

The Report presents a "Catalogue of Congestion Management Measures" contained in Chapter II. Nearly 40 conventional and innovative congestion management measures are presented within nine sections representing nine strategy classes, as follows: Land-use and Zoning, Telecommunications Substitutes, Traveller Information Services, Economic Measures, Administrative Measures, Road Traffic Operations, Preferential Treatment, Public Transport Operations, and Freight Movements. Within each section, the strategy class along with the associated measures are described, its objectives and major impacts stated, the application and institutional responsibilities are discussed, and any special or associated problems presented. Examples of the application of the measures are presented at the end of each section.

Chapter III presents an overview of the policies, plans, and programmes that presently exist and that will be needed in order to enable congestion management measures to be effectively implemented. Case studies are presented in this chapter as evidence of current experience. Chapter IV looks at current and potential future societal policies (e.g. the environment, economy, and advanced technologies) to give a picture of how congestion management measures may need to be developed and shaped in order to be more effective.

Group Members prepared technical information from their countries to meet the desired requirements for Chapter II, III and IV. The information was compiled and organised by the respective chapter rapporteurs. Information used to prepared Chapter III was also supplied from the experience presented at the OECD Expert Workshop on *Congestion Management*, held in Barcelona in late-March 1993.

POLICY ADVICE

The following conclusions and recommendations are contained in Chapter V of the report:

♦ Road traffic congestion can be better managed.
♦ Low-cost, conventional measures can be effective.
♦ Pricing techniques can be effective in congestion relief.
♦ Public support of congestion management measures is essential.
♦ Traveller information is important to congestion relief.
♦ Coordination is an essential aspect of congestion management.
♦ Congestion management efforts need to start small then grow.
♦ The private sector has a role in congestion management.
♦ New policies and laws are needed for congestion management.
♦ New technologies will offer tools for congestion management.
♦ Accessibility must be maintained with congestion management.
♦ Evaluations are needed in congestion management.
♦ Training in the practices of congestion management is needed.

CHAPTER I

INTRODUCTION

I.1. BACKGROUND

Road traffic congestion is a significant problem in many parts of the world. Congestion continues to increase, and the conventional approach of the past -- building more roads -- has proven to be difficult for political, economic, social, and environmental reasons. Building new roads has also been found to compound traffic congestion problems by inducing greater unforeseen demands for travel by auto that quickly fill the additional new capacity. The very high cost of road construction and the increasing environmental concerns in many countries have significantly changed the thinking about how best to address the traffic congestion problems.

Congestion is not simply an issue confined to corridors and activity centres of large cities or urban areas. Congestion affects work-trips and the non-work related trips. It affects the movement of people and the flow of goods. Environmentally, congestion significantly reduces air quality in urban areas. In non-urban (rural) areas and inter-city corridors, traffic is disrupted by incidents, maintenance operations, detours, congestion on tourist routes, and other causes.

From the road user's viewpoint, traffic delays are substantial and growing. Rush hour conditions in many urban areas often extend throughout the day. From the employer's viewpoint, congestion takes its toll in lost worker productivity and cost. Speed, reliability, and cost of urban and inter-city freight movements are increasingly affected by congestion.

Significant increases in congestion can be expected. Many countries report expected increases in urban motorway travel of approximately 50 per cent by 2005, and increases in delays of 400 per cent or more if improvements are not made to the current transport system. Traffic safety as a result of traffic congestion also continues to be a prime concern for most countries.

The causes of traffic congestion can be categorized as either recurring or non-recurring. Recurring congestion is predictable delay caused by high volumes of vehicles using the road during the same time periods (i.e. peak commute periods, holiday periods, or special events) and at certain locations (i.e. intersections, interchanges, or toll plazas areas). Major long-term roadway rehabilitation projects also create significant, recurring, congestion during the construction period.

Non-recurring congestion is unpredictable delay generally caused by spontaneous, unplanned occurrences, e.g. traffic accidents and incidents, unforeseen special events, or emergency maintenance.

13

In most countries weather conditions can also create significant non-recurring congestion problems. Unexpected traffic delays and accidents occurring in snow, rain, or fog conditions are key elements in the development of congestion reduction policies.

Traffic congestion on a city street (Paris)

Given the growing body of evidence to show that traffic congestion problems cannot be solved simply by expanding the road infrastructure alone, many countries are doing more to manage their existing transport systems, in order to enhance mobility and safety, reduce demand for car use, and improve traffic flows. Experience and technical research have demonstrated that when properly applied, measures taken to manage the existing transport system can have a profound impact on trip making behaviour and traffic congestion. A variety of measures and technologies for motorised and non-motorised transport as well as economic and administrative policies have been used to manage congestion or influence travel demand. Many experts in the transport profession also believe that advancements in new technologies will be applied to address congestion management in the future.

I.2. DEFINITION AND SCOPE

Congestion management is the application of administrative, economic, operations, and technological measures aimed at making the most efficient use of existing transport infrastructure, modes, and services. The measures are designed to improve the operating efficiency in three ways. First, congestion management offers the opportunity to increase the use of alternative transport modes including public transport, carpooling, and bicycle/walking. Second, congestion management offers the opportunity to alter trip patterns through the application of measures like land use policies, alternative work schedule arrangements, telecommuting, and pricing. Third, congestion management offers the opportunity to improve traffic flow through measures like traffic signal improvements, incident management and route guidance systems.

Implemented individually or in concert with one another, these measures can help to achieve one or more of eight clearly measurable objectives. These objectives, which can also be considered as positive impacts, are listed as follows:

1. reducing the need to make a trip,
2. reducing the length of a trip,
3. promoting non-motorised transport,
4. promoting public transport,
5. promoting carpooling,
6. shifting peak-hour travel,
7. shifting travel from congested locations, and
8. reducing traffic delays.

Implementing cost-effective congestion management measures becomes especially important if urban centres are to maintain current levels of mobility, increase accessibility, improve safety, and relieve traffic congestion in the face of growing demands for travel, limitations on system capacities and expansion, and environmental constraints. Mobility, and how to maintain it, becomes particularly critical when traffic on many urban roads is estimated to increase nearly 50 per cent by the year 2005, while the road supply is expected to increase about six per cent by the year 2005.

Congestion management measures can be considered throughout a metropolitan area to address system, corridor, individual facility, and site improvements. They can also be implemented along major intercity corridors to relieve traffic congestion problems caused by work travel, construction, holiday or vacation travel, freight movement, and weather.

I.3. DEMAND-SIDE AND SUPPLY-SIDE MEASURES

Congestion management measures are generally implemented to address problems in cities, corridors and activity centres. Corridors include the radial and circumferential travelways within an urban area or between cities. These travelways generally include motorways, freeways, expressways, arterials, and public transport lines. Activity centres are the employment, retail, commercial, educational, special event, and/or recreational areas that generate (or attract) trips (e.g. a suburban office park, a shopping centre, a central business district, or a historic centre).

Congestion management measures are usually considered as "demand-side" or "supply-side" types of strategies. Demand-side measures are designed to reduce car demand on the system by increasing vehicle occupancy, increasing public transport mode share, reducing the need to travel during a specified peak time period, and/or reducing the need to travel to a specified location. These measures include land-use policies, traveller information services, economic and administrative policies, and telecommunications substitutes.

Supply-side measures are intended to increase existing capacity of the system in order to improve the traffic flow for all modes. These measures include roadway operations, preferential treatments, public transport operations, and freight movement.

It is important to note that some demand-side and supply-side measures are related and act to support each other's purposes. For example, supply-side measures to provide preferential bus lanes support programmes to reduce car demand by increasing the effectiveness of public transport measures. Demand-side measures like congestion or parking pricing help to reduce car demand and support supply-side measures for roadway operations and public transport.

Both demand-side and supply-side measures, when applied in an effective manner, by transport professionals hold the potential for improving the management of congestion problems and, in many cases, can result in improvements to traffic conditions. Table I.1 presents the supply-side and demand-side measures, categorised within nine strategy classes that are considered to represent the broad spectrum of congestion management methods. Each of the nine strategy classes and associated measures listed in Table I.1 are described and illustrated in greater detail in Chapter II. Freight and urban goods movement measures are included in the presentation of Chapter II only to the extent that they address congestion management.

I.4. OBJECTIVES AND STRUCTURE OF REPORT

This Report was prepared by the OECD Scientific Expert Group on Congestion Control and Demand Management. It is designed to present the "State-of-the-Practice" and future trends with congestion management measures and strategies.

The Report has three purposes: First, the report presents a catalogue of supply-side and demand-side congestion management measures that incorporates current experience and examples of actual applications. Second, the report addresses policy and institutional factors that are critical to implementing effective congestion management measures, applied either individually or as part of a comprehensive programme. A significant contribution of materials was given during the OECD Expert Workshop on *Congestion Management*, held in Barcelona on 29-30 March 1993, (see Annex A). Third, the report discusses policies, trends, and technological developments that will undoubtedly influence the manner in which congestion management is applied in the near future.

In collecting and preparing the material for this report, the Study Group was guided by the following three criteria:

1.   The emphasis of this study is on actions that can be implemented in the short-term to medium-term, that is 1 to 3 years.

2.   The measures reviewed do not include major construction additions to road supply.

3.   The congestion management measures presented in the report represent projects, programmes, and policies, that use technology currently available. **IVHS and other advanced technology concepts are not discussed specifically in the report.**

It must also be noted that the measures described represent the experience of the countries participating in and contributing to the Study Group. It was recognized that each country has a different approach, depending on circumstances and overall transport policy. Bringing together these individual experiences was a major effort of the Study Group.

The Report is structured to meet the above stated purposes, within the established framework, in the following manner:

Chapter II presents the "State-of-the-Practice" in congestion management as identified from the country sources used in the preparation of the report. The chapter presents this information in tabular form and through a catalogue. The possible applications and impacts of the congestion management measures are presented in tables. The catalogue of measures is divided into nine sections that represent the nine strategy classes appearing in Table I.1. Within each section, the strategy class is described, its objectives and major impacts are stated, the application and institutional responsibilities are discussed, and any special problems associated with the measures are presented. The presentation is further supported by case study examples of each congestion management measure addressed within a strategy class. The examples are based on the contributions prepared by the Expert Group Members. The measures presented and the accompanying examples represent only a sample of the congestion management activities that are taking place. The chapter is not intended to present an exhaustive listing of activities.

Chapter III presents a discussion of comprehensive congestion management programmes and policies that enable congestion management measures to be effectively implemented. The discussion draws experience from the Expert Workshop on *Congestion Management*, held in Barcelona on 29-30 March 1993. The chapter identifies the strategies, techniques and institutional features that have shown to be essential for effective congestion management. Case examples are presented to support the discussion.

Chapter IV presents a discussion on the future of congestion management. The chapter discusses the evolution of transport policies (i.e. with regard to the environment and the economy) and technology that will affect the way congestion management measures are implemented into the future. The chapter also discusses trends and research in the field of congestion management measures as a part of a sustainable development policythat will likely influence the transport sector.

The report concludes with Chapter V which presents the overall conclusions and recommendations that have been derived from the experience and work of the Study Group.

Table I.1. **Classification of congestion management measures**

TYPES OF MEASURE	STRATEGY CLASS	MEASURES
DEMAND - SIDE	LAND-USE & ZONING	* Land-use & Zoning Policy * Site Amenities & Design
	COMMUNICATIONS SUBSTITUTES	* Telecommuting * Tele-Conferencing * Tele-Shopping
	TRAVELLER INFORMATION SERVICES	* Pre-Trip Travel Information * Regional Rideshare Matching
	ECONOMIC MEASURES	* Congestion Pricing * Parking Pricing * Transportation Allowances * Transit & Rideshare Financial Incentives * Public Transport Pass Programmes * Innovative Financing
	ADMINISTRATIVE MEASURES	* Transportation Partnerships * Trip Reduction Ordinances and Regulation * Alternative Work Schedules * Auto Restricted Zones * Parking Management
SUPPLY - SIDE	ROAD TRAFFIC OPERATION	* Entrance Ramp Controls * Traveller Information Systems * Traffic Signalisation Improvements * Motorway Traffic Management * Incident Management * Traffic Control at Construction Sites
	PREFERENTIAL TREATMENT	* Bus Lanes * Carpool Lanes * Bicycle & Pedestrian Facilities * Traffic Signal Pre-Emption
	PUBLIC TRANSPORT OPERATIONS	* Express Bus Services * Park & Ride Facilities * Service Improvements * Public Transport Image * High Capacity Public Transport Vehicles
	FREIGHT MOVEMENTS	* Urban * Inter-city

CHAPTER II

CONGESTION MANAGEMENT MEASURES

II.1. OVERVIEW OF APPLICATIONS AND IMPACTS

A wide variety of congestion management measures have been applied to provide mobility and travel choices for efficiently moving both people and freight. The broad spectrum of information and experience gathered for this OECD report has shown that when properly applied, congestion management measures can have significant positive effects on public transport demand, telecommuting services, non-motorized transport, car pooling activities, traffic congestion, and traffic flow control.

This chapter presents congestion management measures in nine categories -- identified as strategy classes in Table I.1 -- that are being used in the countries researched for this report. The material used to formulate the chapter was obtained from information provided by the OECD Study Group as well as literature provided by OECD Member countries. The information and experience is presented in two different formats - tabular in Tables II.1 - II.4 and in the form of a detailed catalogue (Sections II.2 and II.10).

Two sets of tables illustrate the applications and impacts of the congestion management measures. The first set of tables -- Table II.1 for the demand-side measures and Table II.2 for the supply-side measures -- illustrates typical applications. The tables indicate the relationship of each congestion management measure to each of typical applications of measures identified by the study group. These application types are partly overlapping; however, they portray the broad range in which congestion management measures can be applied.

The application types are as follows:

1. Urban Area: Applied within a city and/or a metropolitan area to address recurring or non-recurring congestion.

2. Inter-Urban Area: Applied in a non-urban setting or an inter-city corridor to relieve recurring or non-recurring congestion.

3. Peak Periods: Applied to manage/relieve recurring congestion during peak commuting travel times in either urban or inter-urban settings.

4. Off-Peak Periods: Applied to manage/relieve non-recurring congestion or improve mobility during non-peak travel periods in either urban or rural setting.

5. Holiday Periods: Applied to manage/relieve congestion during peak non-commuting (non-work) travel times in inter-city corridors or at major activity centres.

6. Construction/Maintenance: Applied to relieve congestion and provide mobility in a corridor where major long-term facility reconstruction or short-term maintenance are taking place.

7. Special Events: Applied to relieve localised congestion created by special events, such as sporting events, festivals, and parades.

8. Incidents: Applied to relieve delays and congestion created by accidents, emergencies and breakdowns on urban or inter-urban roads.

The second set of tables -- Table II.3 for demand-side measures and Table II.4 for supply-side measures -- presents the potential objectives, or positive impacts, of congestion management measures on travel and trip-making. The potential impacts/objectives are as follows:

1. Reduce Number of Trips: As a result of the measure, a trip no longer needs to be made.

2. Reduce Length of Trips: As a result of the measure, a trip can be made in a shorter distance.

3. Promote Non-Motorized Transport: As a result of the measure, greater use is made of non-motorized forms of travel, such as walking and bicycling.

4. Promote Public Transport: As a result of the measure, greater use is made of public transport by those who formerly used the auto.

5. Promote Carpooling: As a result of the measure, greater use is made of carpooling (also know as ridesharing) by those who formerly used the auto alone.

6. Shift Peak-Hour Travel: As a result of the measure, trips migrate to less congested time periods with the original peak period becoming longer, but less concentrated.

7. Shift From Congested Locations: As a result of the measure, travel is diverted away from congested locations.

8. Reduce Traffic Delays: As a result of the measure, travel by auto, truck, and bus is faster, safer, and more efficient.

Table II.1. **Application of demand-side congestion management measures**

TYPES	STRATEGY CLASS	MEASURES	APPLICATION OF MEASURES							
			URBAN	INTER URBAN	PEAK	OFF-PEAK	HOLIDAY	CONST/MAINT.	SPECIAL EVENTS	INCIDENT MANAGEMENT
D E M A N D - S I D E	LAND USE & ZONING	Land Use & Zoning Policy	XX	X	XX	X			XX	
		Site Amenities & Design	XX		XX	X			XX	
	TELE-COMMUNICATIONS SUBSTITUTES	Telecommuting	XX	X	XX	X		X		
		Tele-Conferencing	X	XX	X	X				
		Tele-Shopping	X		X	X				
	TRAVELLER INFO SERVICES	Pre-Trip Travel Information	XX	XX	XX	X	XX	XX	XX	XX
		Regional Rideshare Matching	XX		XX			X	X	
	ECONOMIC MEASURES	Congestion Pricing	XX	XX	XX	X	XX	X	XX	
		Parking Pricing	XX		XX	X			XX	
		Transport Allowances	XX		XX					
		Transit & Rideshare Financial Incentives	XX		XX	X			X	
		Public Transport Pass Programme	XX		XX	X			X	
		Innovative Financing	XX		XX					
		Transp. Partnerships	XX		XX					
	ADMINISTRATIVE MEASURES	Trip Reduction Ordinances & Regulations	XX		XX					
		Alternative Work Schedules	XX		XX					
		Auto Restricted Zone	XX		XX	XX			XX	
		Parking Management	XX		XX	XX			XX	

Note: XX - Significant Application, X - Some Application, (Blank) - No Application

Table II.2. Application of supply-side congestion management measures

TYPES	STRATEGY CLASS	MEASURES	APPLICATION OF MEASURES							
			URBAN	INTER URBAN	PEAK	OFF-PEAK	HOLIDAY	CONST/ MAINT.	SPECIAL EVENTS	INCIDENT MANAGEMENT
S U P P L Y - S I D E	ROAD TRAFFIC OPERATIONS	Entrance Ramp Controls	XX	X	XX	X	X	X	X	X
		Traveller Information Systems	XX	XX	XX	X	X	XX	XX	XX
		Traffic Signalization Improvements	XX		XX	XX		X	X	X
		Motorway Traffic Management	XX	X	XX	X	X	X	X	X
		Incident Management	XX	XX	XX	XX	X	XX	X	XX
		Traffic Control at Construction Sites	XX	X	XX	XX	X	XX	X	
	PREFERENTIAL TREATMENT	Bus Lanes	XX		XX	XX		X	X	
		Carpool Lanes	XX		XX	X		X		
		Bicycle & Pedestrian Facilities	XX		X	X			X	
		Traffic Signal Pre-Emption	XX		XX	X			X	
	PUBLIC TRANSPORT OPERATIONS	Express Bus Services	XX	X	XX	X		XX	X	
		Park & Ride Facilities	XX	X	XX	X		XX	X	
		Service Improvements	XX	X	XX	XX	X	XX	XX	X
		Public Transport Image	XX	X	XX	X		XX	X	
		High Capacity Public Transport Vehicles	XX	X	XX	X		XX	XX	
	FREIGHT MOVEMENTS	Urban Goods Movement	XX		XX	XX				X
		Inter-city Goods Movement	X	XX	XX	XX		X	X	X

Note: xx - Significant Application, x - Some Application, (Blank) - No Application

Table II.3. Potential impacts (objective) of demand-side congestion management measures

TYPES	STRATEGY CLASS	MEASURES	IMPACTS							
			REDUCE NEED TO MAKE TRIP	REDUCE LENGTH OF TRIP	PROMOTE NON-MOTORIZED TRANSPORT	PROMOTE PUBLIC TRANSPORT	PROMOTE CARPOOLING	SHIFT PEAK HOUR TRAVEL	SHIFT TRIPS AWAY FROM CONGESTED LOCATIONS	REDUCE TRAFFIC/ TRAVELLER DELAYS
DEMAND - SIDE	LAND USE & ZONING	Land Use & Zoning Policy	X	XX	XX	XX	X	X	X	X
		Site Amenities & Design	XX	XX	XX	XX	XX	X		X
	TELE-COMMUNICATIONS SUBSTITUTES	Telecommuting	XX	XX	X	X			X	X
		Tele-Conferencing	XX	XX						
		Tele-Shopping	XX	X						
	TRAVELLER INFO SERVICES	Pre-Trip Travel Information	X	X	X	XX	XX	XX	XX	X
		Regional Rideshare Matching	XX			XX	XX	X	X	X
	ECONOMIC MEASURES	Congestion Pricing	X	X	X	XX	XX	XX	XX	XX
		Parking Pricing	X	X	X	XX	XX	XX	XX	X
		Transportation Allowances			X	XX	X	X		X
		Transit & Rideshare Financial Incentives				XX	XX	X	X	X
		Public Transport Pass Programme				XX		X		X
		Innovative Financing				XX	XX			X
		Transp. Partnerships			X	XX	XX	X	X	X
	ADMINISTRATIVE MEASURES	Trip Reduction Ordinances & Regulations	XX		X	XX	XX	X	X	XX
		Alternative Work Schedules	X			XX	XX	XX		X
		Auto Restricted Zone	XX		XX	XX	XX		XX	
		Parking Management	X	X	X	XX	XX	X	X	X

Note: XX - Significant Impact, X - Some Impact, (Blank) - No Impact

23

Table II.4. **Potential impacts (objectives) of supply-side congestion management measures**

TYPES	STRATEGY CLASS	MEASURES	REDUCE NEED TO MAKE TRIP	REDUCE LENGTH OF TRIPS	PROMOTE NON-MOTORIZED TRANSPORT	PROMOTE PUBLIC TRANSPORT	PROMOTE CARPOOLING	SHIFT PEAK HOUR TRAVEL	SHIFT TRIPS AWAY FROM CONGESTED LOCATIONS	REDUCE TRAFFIC/ TRAVELLER DELAYS
S U P P L Y - S I D E	ROAD TRAFFIC OPERATIONS	Entrance Ramp Controls				XX	XX	X	XX	XX
		Traveller Information Systems		X		X	X	XX	XX	XX
		Traffic Signalization Improvements				X	X		X	XX
		Motorway Traffic Management		X		X		X	X	XX
		Incident Management					X		XX	XX
		Traffic Control at Construction Sites				X		X	XX	XX
	PREFERENTIAL TREATMENT	Bus Lanes				XX		X	X	X
		Carpool Lanes				X	XX	X	X	X
		Bicycle & Pedestrian Facilities			XX					
		Traffic Signal Pre-Emption				XX				X
	PUBLIC TRANSPORT OPERATIONS	Express Bus Services				XX				X
		Park & Ride Facilities				XX	XX		X	X
		Service Improvements				XX				X
		Public Transport Image				XX				X
		High Capacity Public Transport Vehicles				XX				X
	FREIGHT MOVEMENTS	Urban Goods Movement						X	X	XX
		Inter-city Goods Movement	X					X	X	XX

Note: xx - Significant Impact, x - Some Impact, (Blank) - No Impact

24

The following sections provide a catalogue of potential congestion management measures which have been implemented and have been proven effective in practical applications throughout the world. The following strategy classes are examined:

- ◆ II.2 - Land Use and Zoning
- ◆ II.3 - Telecommunications Substitutes
- ◆ II.4 - Traveller Information Services
- ◆ II.5 - Economic Measures
- ◆ II.6 - Administrative Measures
- ◆ II.7 - Road Traffic Operations
- ◆ II.8 - Preferential Treatment
- ◆ II.9 - Public Transport Operations
- ◆ II.10 - Freight Movements

For each of the nine congestion management classes of measures, the following information is provided, as appropriate:

- ◆ Description -- gives in general terms the scope of each congestion management strategy class, provides a listing of individual measures which fall within each class, and outlines its applicability to congestion management

- ◆ Objectives and Major Impacts -- describes specific objectives associated with individual measures within that strategy class and summarises observed impacts associated with each measure where available

- ◆ Application of Measure -- presents typical, successful applications for the individual measures with brief, illustrative examples

- ◆ Institutional Responsibility for Implementation -- indicates the most likely implementing parties for individual measures

- ◆ Effects on Travel Patterns -- where data has been collected, describes the quantified impacts of measures on travel patterns in terms of modal split, frequency of trips, etc.

- ◆ Cost-Effectiveness -- presents the results of field observations, or estimates, of costs and benefits associated with the measures

- ◆ Special Problems or Issues -- outlines key issues which must be addressed if the programme is to be successful

- ◆ Examples -- presents selected, operational examples of the measures which illustrate experience and application.

II.2. LAND USE AND ZONING

Description - Land use planning is typically thought of as the formulation and implementation of public policy regarding land use and development. It attempts to create a rational order for the built environment, through both the framework of government regulation and physical design to identify residential, commercial, industrial and other (public, private) areas and to promote environmental protection, public health and safety, and general public welfare. Land use planning can operate on all scales of analysis, from the site-specific to national. The land use and zoning measures which could be part of an overall congestion management programme include comprehensive or master planning, formulation of zoning and other land use regulations (such as site design, construction, and other development performance standards), capital improvements planning and public finance policy, and governmental review of development proposals.

Objectives and Major Impacts - Urban and suburban patterns of land use and the intricacies of individual site designs are both key elements in any effort to manage transport demand and congestion. Auto-oriented patterns of residential or commercial development impose considerable constraints on the potential for success of other congestion management strategies, like public transport operations or preferential treatment for carpools. In contrast, transit[1]-oriented neighbourhoods or employment sites directly enable the other complementary congestion management measures to work to their full potential. The key difficulty associated with land use and zoning measures is that they are long-term measures requiring a consistent, long-term commitment.

On a regional scale, appropriate measures will in the long-term establish land uses and urban patterns that support public transport, bicycling, and walking. These measures will also involve the development of land use patterns that facilitate multi-purpose trips and minimise the number and length of vehicle-trips.

The principal impact of the land use class of measures is the removal of some impediments to the use of public transport, bicycling, or walking. For example, transit/rideshare-friendly work site design eases worksite access for transit/rideshare commuters by removing physical barriers to transit/ridesharing and can provide time savings to transit users if walking distance to the transit stops are reduced.

Many employment sites, particularly in the U.S. (even downtown sites near public transport services), have in the past been designed to serve the motorist commuter first, then the transit commuter. Because site design can directly affect a person's mode choice, site amenities and design attributes can be critical elements in a demand management programme. On-site services minimize commuters' real or perceived need for the use of an automobile for business-related purposes before, during or after the work day.

Application of Measure - In general, effective land use and zoning measures are comprised of elements of the following:

- ◆ Incorporate mixed, compatible land uses;

- ◆ Create neighbourhood commercial districts (e.g. places of work near or in residential areas);

- ◆ Encourage development of recreational, employment, and retail land uses near residential areas;

- ◆ Encourage public transport-compatible development on vacant parcels in developed areas near bus routes and stops;

---

[1]   Transit (American terminology) is the same as public transport.

- Discourage auto-oriented uses near transit stops;

- Increase residential densities along bus routes and at bus stops;

- Increase employment densities in activity centres;

- Explicitly plan for pedestrian and bicycle access to activity centres.

Transit-friendly building sites are those that accommodate the space and manoeuvring requirements of public transport vehicles and vans; provide safe, attractive, and protected loading areas; and minimise the walking distance for transit/rideshare users. At work sites, the on-site services which contribute to the satisfaction of midday needs of a worker include cafeterias and restaurants, dry cleaners, automated teller machines and banks, convenience shopping, grocery stores, video rental stores, and printing and copy shops.

Institutional Responsibility for Implementation - The local zoning ordinance is the primary tool used to implement land use policy. Unfortunately, issues dealing with public transport, bicycling, and walking are seldom addressed in contemporary zoning ordinances. In order to encourage, and in some cases enable, land use planning and design which is sensitive to the needs of public transport patrons, bicyclists, and pedestrians, the ordinances need to be reviewed and updated, as appropriate.

Special Problems or Issues - An effective land use pattern which complements travel demand management objectives requires a long-term and consistent commitment by the public sector. Within

Bicycle-only streets integrated into residential area (Houten, The Netherlands)

the UK in the 1950's, a series of New Towns was designated, mainly in a ring around London as well as two in central Scotland. Among these was Milton Keynes, some 55 km northwest of London. This was considered a linear city because it is situated astride a rail line, a rail transit link, and the A6 trunk road. As cars became increasingly affordable and fashionable, a more dispersed layout of the residential development was permitted to occur. The opportunity for a transit-friendly development was lost. There was only a limited concentration of residences and other activities within walking distance of corridors which could command a viable density of public transport operations.

In order to facilitate pedestrian movement, many communities have reduced road widths, shortened intersection curb radii, and encouraged on-street parking (which, in combination, can lead to a more pedestrian-friendly, and sometimes transit-friendly, setting). Implementation of these actions must be carefully balanced with the mobility and safety needs of motorists.

Examples

Land Use and Zoning Policies

Land use and zoning policies specify the need to designate transit corridor districts or transit service zones; the need to explicitly plan for pedestrian and bicycle access to transit; and the need to provide roadways which can accommodate transit.

Federal Town and Country Planning Law (Switzerland) - The law attempts to create places of work near or in residential areas by using zonal assignment and planning techniques. One of its objectives is to achieve a general reduction in home-to-work trips by automobile. In many applications, the law has resulted in the reduction of city centre parking supply for automobiles, with a particular emphasis on reducing long-term parking for commuters. In combination with other measures (e.g., public transport enhancements and fare reductions), the net effect in the Berne region, for example, is a 40 per cent public transport mode share (70 per cent in the city centre).

The Netherlands reports that since 1990 the Ministry for Housing, Regional Development and the Environment has encouraged businesses to locate near public transport services. An evaluation of the impact of a business relocating near public transport services in The Hague found a 50 per cent increase in employee usage of public transport, a 50 per cent decrease in the use of cars, and the use of bicycling decreased 25 per cent.

In Melbourne (Australia), future urban growth is restricted to a limited number of localities or corridors, and within these, growth is focused upon public transport routes serving both local travel within the corridor (mainly buses), and long distance radial travel to the central city.

Traffic Planning Guidelines (Sweden) - There is a general declaration in the Traffic Planning Guidelines that building development should be located and laid out in such a way that it minimises the need to travel by automobile and favours the use of public transport.

Selected US reference

U.S. DEPARTMENT OF TRANSPORTATION (1991). *Guidelines for Transit-Sensitive Land Use Design.* Washington D.C.

II.3. TELECOMMUNICATIONS SUBSTITUTES

Description - The use of telecommunications technologies as a substitute for motorized trips holds the potential to reduce congestion. Telecommuting, teleconferencing, and teleshopping are all forms of telecommunications as a substitute for travel. Telecommuting (also called telework) is the partial or total substitution of telecommunications (with or without the assistance of computers) for the daily commute to/from work. Teleconferencing is the substitution of television and telephone communication (i.e., audio and/or video) for trips taken to meet directly with several individuals or groups and is most often used for business purposes. Teleshopping involves the use of the telephone to shop and purchase items without physically travelling to a store.

All three types of telecommunications substitute measures are in their infancy in terms of widespread application. Potential regionwide and site-specific benefits and costs can only be speculated from the limited experience.

Objectives and Major Impacts - Telecommuting potentially affects both employees and employers. Telecommuters may make fewer work trips or fewer and shorter trips, depending on whether they telecommute from home or from a satellite work centre. Telecommuting also may influence mode choice. Telecommuters may switch from solo driving to walking, cycling, or transit to access neighbourhood or satellite centres which are closer to home than their main office. Indeed, they may also switch from carpools and transit to solo driving or cycling as a result of increased work flexibility.

Telecommuting may affect non-work trips. For instance, telecommuters or their family members may make more midday shopping trips as a result of having flexibility in work time or a vehicle normally parked at work. Telecommuting also may affect commuter decisions about where to live (i.e., they may choose to live even farther away from the city centre).

Teleconferencing will have in the future a more noticeable impact on business travel. It is expected that teleshopping services will have only a small effect on travel patterns.

Application of Measure - Telecommuting is an approach for reducing home-to-work trips by allowing employees to work at remote locations from the traditional office (at home, or at worksites located closer to home than to the office). Employees may be linked to the work place by computer and modem, or simply by telephone. Remote location options include employees working at satellite work centres (run by single employers) or at neighbourhood work centres (run by multiple employers).

Teleconferencing can substitute for some per cent of business travel. Teleshopping has been commonplace for many years. However, there have been several recent applications of telecommunications technology for the purposes of teleshopping. One such application is the use of Minitel devices in France where consumers type their orders to be delivered the next day. The delivery services are privately provided, but the telephone and Minitel services are provided by the France Telecom public agency.

Institutional Responsibility for Implementation - Telecommuting is most often implemented by employers, and does not require a substantial investment for implementation. Some local governments are experimenting with the concept in conjunction with private sector parties (in particular, the telecommunication, banking, and hardware and software development industries).

Teleconferencing is most often implemented by private businesses. It requires a substantial investment in equipment (e.g. studio, transmission lines) or the renting of facilities which have the equipment available.

Teleshopping is strictly a private, commercial venture.

Effects on Travel Patterns - Travel effects of telecommuting have not been established beyond theoretical scenarios and limited sample/time case studies. All investigations of telecommuting impacts suggest significant reductions in work trip vehicle usage. In one ongoing telecommuting programme in Los Angeles, not only were travel impacts positive, but impacts on productivity were also positive. Management raised few concerns about lack of availability for meetings and communications with staff.

Another study of telecommuters by the State of California (USA) found that work trip rates decreased 30 per cent when the workers telecommuted between one and two days per week.

A satellite telecommuting demonstration project found that 93 per cent of employees reported a reduced number of work trips. Travel time savings were over 7 hours per week (or 385 hours per year).

Cost-Effectiveness - Because telecommuting is a relatively new strategy, it is difficult to estimate potential benefits. Much depends on the future growth of telecommuting, the mix of industries in the future and unforeseen technology advances. One forecaster estimates $23 billion could be saved annually in transport, environmental and energy costs if there is a 10 to 20 per cent increase in activities done through telecommuting instead of physical transport. These telecommuting benefits compare quite favourably to the benefits attributed to other congestion management measures described in this chapter and to those estimated for future deployment of advanced technology programmes such as Intelligent Vehicle Highway Systems (IVHS).

Special Problems or Issues - Telecommuting is an important element of employer demand management programmes. Not only can the strategy reduce the number of work trips for those working at home, or reduce the length of the commute for those working in satellite centres; it may dovetail with other employer objectives including improved morale and productivity. In a pilot project in Los Angeles, 80 per cent of telecommuters reported an increase in work productivity, and the majority of supervisors said productivity increased. However, it should be noted that other non-work trips (e.g., leisure) may be created as a result of telecommuting.

Telecommuting workstation

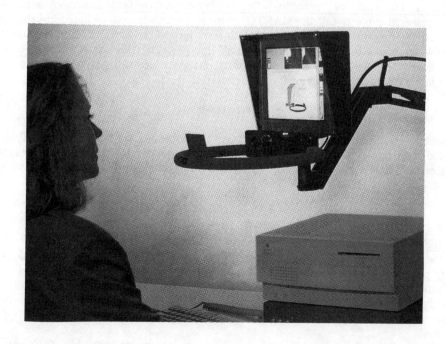

Institutional and societal barriers appear to pose the most significant obstacles to telecommuting. Management attitudes (e.g. of the need for face-to-face contact), security issues, contractual relationships, liability issues, health and safety, and zoning restrictions are some of the key obstacles. These barriers, however, can be overcome through education, training, and changes in laws. New skills involved in telework have yet to be institutionalised.

Examples

Telecommuting

Southern California Association of Governments (SCAG) (Los Angeles, USA) - In a test of employees at SCAG, fourteen per cent of employees at the agency participated in the experiment. Average participation was once every nine days. Most telecommuters worked from home. One worked at a satellite work centre. Not only were reported travel impacts positive, but impacts on productivity were positive. Management raised few concerns about lack of availability for meetings and communications with staff.

State of California (USA) - In preliminary findings among State of California telecommuters, work trip rates decreased 30 per cent from 0.9 trips per day to 0.63, compared to a control group where work trip rates did not change. In the project, workers telecommuted either one or two days per week. The project involved over 400 State employees across 13 agencies. Travel diaries were used to track travel impacts.

Ministry of Transport and Public Works (Netherlands) - The Ministry tried two experiments with reducing the need for auto trips by telecommuting. The objective of the experiments was to reduce cars travelling in rush hours by five per cent. The teleworking measure was expected to cause changes in mode split, time of departure, number of travel days, and number of journeys. The participants worked 20 to 60 per cent of their normal work hours at home, using a personal computer, modem, printer, telefax, and an extra telephone connection. The main observation was that when an average of 20 per cent of the teleworking time can be planned with flexibility, then 15 per cent less commuter journeys are made. There was no increase in car use by other family members. Other effects are not yet reported.

The Federal University of Technology (ETH) (Zurich, Switzerland) undertook a major congestion management research project, called "MANTO". Its recommendations included suitable utilisation of telecommunications for work and shopping. A follow-on project called Telecommunication Communities (KMG) is now attempting to put the earlier proposals into practice, with varying degrees of success. The number and density of telecommunications resources in service are still too low for it to be possible to measure the influence on the development of congestion.

Teleconferencing

No examples of teleconference impacts on travel or pilot tests have been reported. Teleconferencing was advocated in the late 1960's in the United Kingdom on the grounds of business travel reduction benefits and was linked to a large-scale effort toward office decentralisation from London. However, there is only limited documentation of the effort.

Teleshopping

Minitel Services (France) - One of the services offered by Minitel (primarily a videotext information service) is for consumers to type orders for goods to be delivered the next day. The teledelivery services are privately provided, but the telephone and Minitel services are provided by France Telecom (a public agency). In the 1992 Minitel directory, 17 servers represent catalogues, 30 represent neighbourhood retail, 88 represent food items, and 184 represent miscellaneous items. Banking services, and train or air booking services also exist on Minitel.

CompuServe (U.S.), Prodigy (U.S.) and other such services in use in the UK, Germany, France, etc. are available for ordering a variety of goods or services. Most services are offered on a subscription basis and require the use of a modem by the consumer.

U.S. DEPARTMENT OF TRANSPORTATION (1993). *Transportation Implications of Telecommuting*. Washington DC.

II.4. TRAVELLER INFORMATION SERVICES

Description - Traveller information services are programmes that provide information on routes to take and/or time periods to use in order to avoid congestion (via the print media, radio, television, or traveller information centres) and information to travellers on carpooling, transit, and alternative transport modes (e.g. regional rideshare matching programmes). These programmes are offered by many groups and can be targeted to various travel markets including: commuters, students, shoppers, and recreational travellers. They can help the user to make a "permanent" decision (such as location of a residence) or a more "temporary" decision (such as the route or time-of-day for travel). These programmes are structured to help the traveller make appropriate decisions prior to starting the trip; programmes which provide information to the motorist while enroute are described later in Section II.7. **Road Traffic Operations**.

Information can be provided at a range of levels of sophistication and convenience to the potential user. The most basic level is passive postings (such as on carpool ridematch boards, information "take one" displays, transit "hot lines", and roadside signs) that inform travellers of assistance available from a remote source such as the regional public transport operator or regional ridesharing agency. The highest level of information assistance is provided by a traveller information centre, centrally located within a commercial development, at a public transport station, along a major travel corridor, or at an individual employment site. At this level, the traveller receives personalized assistance. These centres are staffed, generally full-time, and provide information on available services, ridematching, and personalized travel planning. They also can serve as outlets for distribution of transit fare media or other information products.

The most basic form of providing information on routes to take in order to reach a destination is a hard-copy map. Any physical means taken to provide information to potential travellers prior to their making a trip would constitute traveller information. The simple advertising of the availability of specialised public transport services, of anticipated traffic congestion associated with special events or with major highway construction, or of routing options during recurring periods of traffic congestion (e.g., immediately prior to and after peak holiday periods) are all traveller information measures which are commonplace in urban areas.

Objectives and Major Impacts - The primary objective of traveller information centres is to make it as convenient as possible for a traveller to access information on available transport services and routes. A traveller information programme may attempt to reduce or remove the perceived advantage of driving alone by making information on other travel options easily accessible and by removing some of the uncertainty and discomfort travellers may feel when trying an unfamiliar travel mode.

Other traveller information programmes encourage travellers to use the road system in an efficient manner (i.e. to use the least congested travel paths and to travel during less congested time periods).

Systematic efforts to monitor the effects of traveller information programmes are underway, particularly under the auspices of IVHS demonstration projects.

Application of Measure - Many of the current applications of traveller information programmes focus on providing information and assistance for commuters. However, the information needs of all travellers must be addressed. An example of one such comprehensive traveller information strategy is the provision of a national telephone number in Italy where a traveller can obtain information on the inter-urban motorway system.

Numerous options exist for disseminating traveller information, depending on the scope and size of the target market:

♦ Regional distribution may include mass mailings; newspaper, radio, and television advertising; and roadside signs, such as those that list "Pool" phone numbers of ridematching agencies. In France, there is a network of road information centres which distribute information on traffic conditions to motorists (at both the local and national level). The road information system is overseen by the Ministry of Transport.

♦ Two of the more comprehensive information services are known as the Bison Futé scheme in France and the Ferienverkehrsprognose (holiday traffic forecast) in Germany. They provide information either several months or several days in advance. The systems utilize historical experience with long-range forecasts. Bison Futé uses all elements of the media (free maps, free calendars, and radio and TV bulletins) to broadcast its message.

♦ At individual employment sites, information dissemination typically relies on bulletin boards, flyers distributed desk-to-desk, in-house newsletters, new employee orientation, and periodic promotional events such as rideshare fairs.

♦ Local area information dissemination utilizes elements of both the regional and individual employment site programmes, for example, mass mailings to new tenants or new homeowners, information distributed through realtors and building managers, posted notices, newsletters, and promotional events. Some information centres are placed in train stations, transit malls, or wherever the general public congregates.

Institutional Responsibility for Implementation - Traveller information programmes typically are sponsored by national or regional transportation agencies, by public transport operators, and by local and regional governments. Traveller information may be targeted to a small audience in a defined local area, for example an employment, shopping, or residential complex. Developers and property managers often are the sponsors of these programmes. Information can also be offered to commuters at individual employment sites. Here, the programme is sponsored by employers, who promote the use of alternative modes to their employees.

Effects on Travel Patterns - Information programmes (i) support mode shifts to alternative means rather than causing the shift and/or (ii) direct a motorist to a preferable route and timeframe. The direct effects of traveller information programmes which encourage alternate routes and trip schedules for travel by automobile have not been well-documented. Generally, the traveller information programme is implemented in conjunction with other congestion management measures such as traffic surveillance and control measures.

Regarding their modal shift impacts, the influence of traveller information programmes generally comes after other demand or supply management measures (or individual circumstances such as lack of access to an automobile) gives the traveller opportunities to choose between different alternatives, other than solo driving. They encourage consideration of alternatives by removing secondary impediments -- real, attitudinal or psychological -- to riding public transport or bicycling, for example.

Research from numerous sources suggests that rideshare matching measures, although an important element of an overall congestion management programme, largely are ineffective alone. One study commissioned by the U.S. Department of Transportation examined the impacts of several factors, among them the presence of an on-site transportation coordinator, on commute mode shares at 46 suburban employment centre sites. It found that sites with a transportation coordinator had a greater than 3 per cent reduction in drive-alone commuting compared to sites without a coordinator.

Cost-Effectiveness - Information programmes accrue expenses for staff and expenses for preparation of marketing materials, programme advertising, and special promotions such as prizes or fairs. Costs vary greatly depending on the situation. The least expensive programme for an individual employer would involve a part-time transport coordinator and minimal programme promotion. The next step would be to have one full-time transport coordinator and more substantial promotion costs. Area-wide programmes, which typically have a larger target market, usually hire more than one staff member and have correspondingly higher costs for both staff and promotional materials.

One estimate (source: U.S. Department of Transportation) of a programme's cost effectiveness found a typical value of less than $7 per day to remove each commute trip from the road. Information programmes generally support the implementation of more tangible programme elements, however, and the total demand management programme trip reduction probably would be higher. With the same level of programme effort, the contribution of information to the total cost per trip reduced would likely be smaller.

Special Problems or Issues - One real impediment addressed by information dissemination is travellers' lack of awareness that options to driving alone exist and that incentives are provided for use of the options. Increasing awareness of such services will lead to an increase in use of alternative modes by travellers who are receptive to a change but need information on ridesharing partners or transit services.

One psychological impediment to alternative mode use is a reluctance to give up the familiar, that is, fear of the unknown. Personal assistance of a trained transport coordinator can ease the transition from car driving to other modes for travellers who might be hesitant about making the shift. A transport coordinator can assist with individual trip planning, describe the new travel situation, screen potential rideshare partners, and facilitate introductions of the ridesharers.

Examples

Pre-Trip Travel Information

Bison Futé Scheme, France - For 15 years, Bison Futé has been providing road users with advice to help them plan their daily or holiday journeys. Bison Futé announces several months or days in advance (up to a year before) and advises about the best departure times and the best diversion routes. Bison Futé uses all information media: free maps indicating congestion points and main diversion routes; free calendars indicating green, orange, red, and even black days for driving in France; and radio and television bulletins during holiday periods. Bison Futé maps carry extensive advertising which must make a significant contribution to cover the cost of the service.

AUTOSTRADE Information Centre (Italy) - AUTOSTRADE is the operator of the largest share of the Italian inter-urban toll motorway system (approximately 2,800 km in length). The AUTOSTRADE Information Centre provides by telephone (twenty-four hours per day) useful pre-programmed and real-time information regarding the AUTOSTRADE network.

Cities of Den Bosch and Eindhoven (The Netherlands) - An information booth was located near the railway station and the adjacent bus station in both cities. The booths have a microcomputer with a touch-screen monitor; a digitized public transport network, road network, and transit timetables; and coordinates of railway stations, bus stops, and key destinations. The measured effect is that 3 per cent of the current transit patrons formerly travelled by another mode. The majority of the information users live more than 25 kilometres from the stations and needed the information. Each automated information desk costs about DFL. 80,000. Advertising possibilities to defray some of this cost are being investigated. Similar computerized public transport information booths (with printing capabilities) have been installed in strategic locations in Paris by the RATP (urban Metro and bus) agency.

Project "Countdown" for London Transport (London, UK) - The London Transport is undertaking a major demonstration project to provide real-time passenger information at bus stops. The system, known as "Countdown" will operate on Bus Route 18 which runs from Sudbury to Baker Street and Kings Cross in London. Passenger research showed that the uncertainty of waiting time is the biggest deterrent to bus travel. A "Countdown" electronic sign will be placed in each major bus shelter along the route and let passengers know exactly how long they will have to wait for the first bus and its destination. Information on the second and third bus will also be provided in case the first bus is not going the entire way or is filled. The electronic sign will also show which other routes serve the stop and can display special messages sent by a Service Controller at the garage about special traffic problems on the route.

The sign is incorporated into bus shelters and at a few locations where shelters cannot be provided. In addition, an audio unit will be fitted into the shelter to "speak" the same information when a button is pressed. Some buses will be fitted with an indicator to show the name of the next stop to passengers which will automatically be updated as the bus proceeds along the route.

The "Countdown" system draws its information from automatic vehicle location (AVL) equipment located along bus route 18, allowing accurate predictions to be made of bus arrival times at the stops fitted with the information displays. The arrival predictions are based on the usual journey times modified by the actual speeds of the three previous buses. This therefore takes account of prevailing traffic conditions. The constant updating of the system helps prevent inaccurate forecasts being made about bus arrivals.

Regional Rideshare Matching

<u>Commuter Transportation Services, Inc. (CTS)</u> is a private non-profit company serving the Los Angeles (USA) area. CTS works primarily through employers' transport coordinators to promote ridesharing among employees, but also assists "unaffiliated" commuters via an on-line telephone matching system. The organisation provides matchlists to employees and "master" lists to coordinators, and processes employee survey data into commute management plans for employers. CTS also provides information on transit service and vanpool vendors. CTS' funding is primarily public (federal, state, county, and City of Los Angeles), but some businesses also contribute.

Approximately 250,000 commuters are registered with CTS and the organisation estimates that nearly 340,000 individuals have been placed into ridesharing arrangements by CTS since its inception in 1974. A comparison of TDM experience for CTS clients with that of non-clients showed that CTS-assisted firms generated 10 per cent fewer trips than non-clients.

<u>Concord Commute Store (Concord, California, USA)</u> - In May 1990, the City of Concord opened the Concord Commute Store, a commuter information centre open to the public in downtown Concord. The City had offered commute information before, but from a less prominent location in the City's Finance Department. The new location was far more visible and accessible. The Store's services include rideshare matching, transit fare sales, transit trip planning, bicycle maps, and road construction updates. Store staff conduct commuter promotions at major employment sites, and actively market services to residents through local newspapers, radio, television, and a City-published newsletter. From June 1990 to June 1991, the Store handled over 2,000 requests, an increase of 150 per cent over the previous year, before the move. The Store's transit pass sales also increased dramatically after the move, by 238 per cent, from $18,071 to $61,137.

<u>Nuclear Regulatory Commission (Montgomery County, Maryland, USA)</u> - The Nuclear Regulatory Commission (NRC) markets TDM to its employees with a staffed, on-site information centre, prominently-located in the employee services office (other services include: travel agent, credit union, etc.). At the centre employees can register for ridematching services, receive transit information, and purchase discounted transit passes. NRC also offers a self-service interactive computer information system just outside the centre and promotes TDM through regular features in the monthly newsletter, and new employee information packets.

<u>Traffic Information Coordination/Electronic Transmittal System (TICETS) (Philadelphia, USA)</u> - The Philadelphia region is developing a programme which will integrate a variety of existing information sources and transmit the information to the travelling public. Data sources will include two traffic advisory services operating in the area, the resident incident management teams, local business organisations, and the operating agencies for the highway and public transport systems. By incorporating the diverse set of information sources, it is the intent of the project to supply real-time information to the widest practical set of travellers needing the information both prior to and during a trip.

Selected references

1. ORSKI, K. (1991). *Evaluating the Effectiveness of Travel Demand Management.* ITE Journal. Washington, DC.

2. FERGUSON, E. (1990). *Transportation Demand Management, Planning, Development and Implementation*. APA Journal. Washington DC.

3. SCHREFFLER, E. and KUZMYAK, J.R. (1990). *Measuring the Impact of TDM Techniques and a Prognosis for Region-Wide Results*. COMSIS Corporation.

II.5. ECONOMIC MEASURES

Description - There are financial incentives and disincentives to bring about changes in transport modes and/or travel periods. Economic measures can take a number of forms, including cash payments, price discounts, or other positive incentives having an economic value. Or, economic measures might be in the form of charges or penalties to discourage certain types of travel, such as travel under highly congested conditions. Peak-period fees for road use or parking fees are examples of such types of charges. Economic measures might be temporary measures intended to encourage first-time use of an alternative mode, or they might be more permanent measures designed to continuously alter the relative costs of alternative transport modes.

The range of potential economic measures, further described in the following sections, includes:

- ◆ Congestion pricing: a user charge to the motorist that accounts for the costs imposed on all motorists as a result of the additional delay caused by that motorist's entry and movement through the traffic stream; these include motorway tolls, point tolls (at a bridge, tunnel, or fixed point on roadway), cordon tolls (a charge for crossing a cordon surrounding congested area), and areawide congestion charges (cordon, or congestion-based by time and distance); motorway and point tolls are the most prevalent applications;

- ◆ Parking pricing: a fee charged for a vehicle parking in a garage, lot, or other parking facility at a worksite, on-street, or at any other location which generally represents a surcharge for peak period arrivals or departures; the fee structure typically favours travel modes other than drive-alone, long-term parking, and off-peak usage (see also Section II.2.5. **Administrative Measures** for a discussion of parking management measures);

- ◆ Transportation Allowances: regular, periodic payments made to all travellers, including those who drive alone, to be used to defray the costs of travel; sometimes set to be equal to a parking charge; employees can buy parking or allowance to other travel costs may be applied;

- ◆ Public transport/ridesharing/non-motorised financial incentives: subsidies which are regular, periodic payments made to travellers who use carpool, public transport, bicycling, or another alternative to driving a car alone; or other indirect financial incentives that have a measurable, monetary value, but which are not direct subsidy payments;

- ◆ Public transport pass programmes: schemes in which a public transport operator offers multiple-ride transit fare media which cover multiple lines and transport modes, generally with a discount from average daily rate;

- ◆ Innovative financing strategies: in most urban areas, the revenue from fare-paying passengers is insufficient to cover public transport operations costs; if economic incentives in the form of lower public transport fares are to be considered, innovative funding sources will need to be investigated.

Objectives and Major Impacts - The objective of economic measures is generally to shift some trips to off-peak travel periods, to public transport, to bicycle usage, to less congested routes, and/or to cause travellers to make more efficient trip decisions. The two key factors in a travel decision are the cost of the trip (both monetary and in terms of time) and the convenience of the trip. Economic measures target the perceived (or real) cost advantages associated with the drive-alone mode of travel by reducing the direct financial cost of alternative modes. Therefore, any tariff structure and its objectives should be publicly understood and accepted by road users.

Application of Measure - Despite clear evidence that economic measures can have significant effects on travel demand, their use is not widespread. This lack of experience is primarily because of their potential cost, and due to political opposition to measures that might impose significant new costs on travellers. As congestion problems worsen, however, economic measures are receiving greater attention, particularly where options for increasing road capacity are limited. Economic measures can be applied on an areawide basis, or they can be structured to apply only to specific corridors, or to key transport bottlenecks. They can be applied only to commute trips, to all trips, or to trips taken under congested conditions. They can also be intended to encourage use of a specific mode, such as public transport.

Congestion Pricing

A form of congestion pricing for roads has been in operations for many years in Singapore which is now considering changing to an electronic metering system within the next few years. In Norway special area entry permits are also being used in Oslo, Bergen, and Trondheim, although their primary purpose is to raise revenues. There is evidence, however, that there has been some impact on overall traffic levels in the controlled area in Bergen. Peak-period entry restrictions in Milan (introduced for environmental reasons) have been successful in reducing road traffic and inducing shifts to public transport. Differential road charges are being successfully used outside Paris on a motorway overloaded. A successful test of electronic road pricing was conducted in Hong Kong in 1986, but the system was never implemented due to political opposition. Congestion pricing options are also being explored for urban areas in several countries, including Germany, United Kingdom, the Netherlands, Sweden, and the United States. In the U.S., there have been comprehensive studies of congestion pricing for the cities of San Francisco and Los Angeles. Whether this interest will translate into support for the implementation of congestion pricing is yet to be seen.

Parking Pricing

Parking charges are also being looked to as a possible means of managing travel demand. Applications of parking charges include parking taxes, parking "cash out" requirements, and ticketed/metered on-street parking. Parking taxes can take several forms (e.g. charges for long-term parking, peak-period parking surcharges) but are generally imposed by a local jurisdiction and are typically undertaken as a revenue raising measure.

Nearly all major cities in industrialized countries have instituted parking pricing and management as part of an effort to manage traffic and space in city centres. In particular, cities have increased parking fees and have increased fines for and surveillance of unauthorized parking. In the most advanced of such programmes, the overall goal is to eliminate "free-of-charge" parking availability in city centres, such as in Switzerland, the Netherlands, and Denmark. Parking fees in several Swiss cities (e.g. Berne, Zurich) were increased significantly -- between 50 and 100 per cent -- in order to reduce the number of persons arriving by private automobile.

Under a parking "cash out" requirement, employers who offer free or subsidized parking to employees must (i) start charging for on-site parking and (ii) offer employees a cash travel allowance equal to the value of the subsidized parking. The objective is to encourage a shift to ridesharing or transit by reducing the demand for parking among those who now park free or at subsidized rates. Driven by air quality concerns, new legislation in California (USA) establishes a parking cash out requirement for large employers in air quality nonattainment areas.

Metered on-street parking is most often used as a revenue-raising measure by local governments, but its use as a traffic management device is becoming more common. In Germany, an objective of all large cities is to provide no free of charge parking in the city centres. Several European cities have installed on-street parking meters from which parkers purchase tickets and place the ticket on the inside of the car window. These programmes have met with success primarily in raising revenue for cities. It should also be noted, however, that if not properly planned, a parking pricing programme that does not provide sufficient inexpensive parking spaces in cities can often contribute to traffic congestion as motorists (especially unfamiliar motorists) search for an acceptable parking space.

Transport Allowances

The typical application of a transport allowance is by an individual employer in the form of a periodic cash payment or as a one-time income adjustment. Institution of the transport allowance programme also typically coincides with the removal of all free parking for employees and the initiation of other compatible travel demand management support measures.

Public Transport/Ridesharing/Non-Motorised Financial Incentives

Another economic measure is the provision of direct financial incentives to use public transport, to carpool, to bicycle, or to walk. One example application is the use of public transport pass subsidies by an employer or developer. In some cases, the employer purchases monthly passes and sells or gives the fare media to employees. Several years ago, the Swiss Federal Administration and other public agencies introduced a parking fee (10 SFr. - 60 SFr. per month) for their employees. As an incentive for employees to use public transport, the Swiss Federal Administration offered a free 1/2-Tax season ticket for public transport service to all employees.

Whereas public transport/rideshare subsidies apply to users of alternative commute modes, travel allowances represent a monthly stipend for employees to use on whatever travel mode they wish, including driving alone. Travel allowances are most often tied to parking charges whereby employees can apply the allowance to all or part of their parking fees or use the allowance to purchase a bus pass or share the cost of commuting in a carpool. For employees who bicycle, walk, or are dropped off at work, the allowance becomes a windfall.

There are several other financial incentives that provide a real, monetary incentive to users of alternative travel modes, but which do not involve direct subsidy payments to users. These include:

- ♦ use of fleet vehicles for ridesharing
- ♦ free or discounted fuel, maintenance and repair for pooling vehicles
- ♦ extra vacation time for users of alternative commute modes
- ♦ use of bicycles

♦ free or discounted equipment (e.g. walking shoes, bicycle helmets)

Vanpool subsidies are another type of employer financial incentives and can take many forms (note: vans are typically 15-passenger vehicles). Employers can provide the vehicles, underwrite insurance and capital costs, etc., or provide direct subsidies to users of vanpools, no matter who owns, leases or operates the vans. Vanpools generally serve long distance commuters, and since vanpool fares are usually distance-based, the monthly fare can be substantial.

Public Transport Pass

In the Paris region the public transport pass programme has been taken a step further through the payback by employers of the "Carte Orange" pass cost to their employees. All employers pay half of the already-subsidised transport pass cost.

Innovative Financing

One successful programme has been the "versement transport" (transport tax) in France which since 1971 has enabled public transport to be partially financed by employers. All urban areas with a population over 100,000 and most of those between 30,000 and 100,000 are taking advantage of "versement de transport." At the current time, the maximum rates are between 0.5 and 2.2 per cent of each employee's salary (paid for by the employer). Another example can be found in Germany where 2 per cent of the fuel taxes are directly used to sponsor public transport projects.

Institutional Responsibility for Implementation - Economic measures can be offered by employers and developers, local governments, public agencies, and others. Economic measures implemented by employers at individual worksites include: rideshare subsidies, transport allowances, subsidised vanpool operation, fee parking, and indirect financial incentives. Employers who have implemented economic measures generally have done so to reduce congestion at their site ; to reduce employees' demand for limited parking; or, as mentioned earlier, to respond to local regulations. Residential and commercial developers also offer similar measures to residents and tenants, again generally to meet traffic mitigation requirements.

Publicly-offered measures generally include subsidized public transport passes and vanpool start-up incentives. These measures, offered by public transport agencies and municipal governments, typically are applied jurisdiction-wide for political reasons, but can be offered in more limited areas, such as a congested highway corridor, to respond to a local traffic or growth problem. In the latter case, it is often necessary to create a dedicated authority.

Pricing measures are typically applied by a local government jurisdiction, but often involve some form of participation by regional or national governments. A private sector road initiative in the U.S. has also proposed congestion pricing as part of its financing plan. Parking cash out programmes could involve regional and local governments, and also include employer participation.

Effects on Travel Patterns - Empirical evidence of the ability of financial incentives to cause intended shifts in travel patterns is quite persuasive. An assessment (conducted for the U.S. Department of Transportation) of over 20 employer-based transportation demand management programmes in the U.S. suggests that financial incentives for the use of commute alternatives are effective in reducing vehicle-trips between 8 and 18 per cent. Financial disincentives in the form of parking charges (even when not coupled with a financial incentive package to use alternative modes) can also produce similar results. When financial incentives are combined with parking charges, the reductions can approach 50

41

per cent (for example, in Enschede, The Netherlands). Area-wide trip reduction has generally been much less dramatic, on the order of 2-5 per cent, due primarily to low overall participation rates.

It must be noted that most successful programmes of economic measures, such as those noted above, include other measures, such as preferential parking for ridesharers, in addition to financial incentives. Thus the trip reduction and mode share changes cannot be credited solely to the economic measure. However, comparisons of programmes with and without economic measures do show, as desired, lower drive-alone rates when they are offered. Recent studies in the U.S. have concluded that rideshare subsidies and parking fees are two of the primary components of effective employer trip reduction programmes. Further, studies also show that a subsidy has approximately the same effect as an equivalent surcharge on driving alone. In other words, a subsidy in a particular monetary amount for ridesharing has an effect similar to that of an equal dollar parking charge for solo drivers. This indicates that positive reinforcing incentives can be as effective as charges and fees that are negatively perceived as penalties.

International experience with congestion pricing for roads, although somewhat limited, provides some valuable lessons. Road pricing has effectively reduced traffic in several cities. In Singapore, the system has resulted in a stable level of trip reduction of 23 per cent. In central Milan the peak-period entry restrictions implemented have led to an estimated 50 per cent reduction in peak period automobile trips into the city centre. A similar system in Bergen, Norway, has reduced traffic in the controlled area by as much as 7 per cent. The Autoroute du Nord congestion pricing scheme near Paris has shifted some 6 per cent of peak-period Sunday return traffic to the Paris region to other time periods.

It is difficult to distinguish the effect of public transport fares and the effect of quality of service, especially in a context when travel demand increases and mobility patterns change. For instance, in the Ile-de-France region, as a result of improvement of suburban/regional public transport (RER services), the introduction of the Carte Orange pass, and the reimbursement by employers, public transport traffic in the region increased between 1975 and 1989 by approximately 34 per cent at the same time that the number of vehicles entering and exiting Paris increased 21 per cent.

Cost-Effectiveness - The cost effectiveness of an economic measure can be expressed as the programme cost divided by the number of vehicle trips reduced at the site or in the area. Programmes that offer financial incentives and those that combine subsidies with parking controls exhibit a broad range of cost effectiveness, depending on the level of the economic incentive and the number of programme participants. Two recent studies of nearly 20 employer programmes in the U.S. showed a cost range between $0.24 and $13.52 per daily trip reduced. However, parking revenue and avoidance of capital costs can greatly reduce the cost per trip reduced, cross-subsidize financial incentives for ridesharers, or even result in a net cost saving.

Information on the cost effectiveness of public agency provided incentives, subsidies and discounts for public transport and vanpooling generally is not reported in terms of cost per trip reduced, but a review of several public, area-wide programmes suggests annual costs of $100 to $200 per participant. Based on the limited number of reports on the effectiveness of economic measures, it appears that public transport pass discounts, when combined with employer subsidies, can contribute to significant trip reductions, with little or no net cost to the public transport provider.

Comprehensive benefit/cost analysis of congestion pricing has not been done, although some indication of potential benefits and costs can be gained from available literature. A more comprehensive assessment of the full benefits and costs of congestion pricing must go beyond the simple comparison of revenues generated and costs of administering toll facilities. A rough assessment for the U.S. concluded that user fees based on marginal costs could raise about $80 billion per year in

1981 (compared to about $40 billion expended on roads by all levels of government, and $22 billion in user fees collected) with a net social gain of about $10 billion.

Special Problems or Issues - Economic measures offer their greatest impact, if they are supported by a comprehensive demand management and system management package. Obviously, the targeted travel mode alternatives (e.g. public transport, bicycle) must be available. But, strong employer promotion and corporate backing also must be present. Financial incentives can be effective only if employees are aware of the incentives and the alternatives and feel management supports their use. It is particularly important to reinforce employees' awareness of the programme over time, otherwise, new employees and those whose travel patterns change might not participate. Just as critical is the support of employers. If they feel that these economic programmes are "social overheads" which make them less competitive, they will be reluctant to participate.

A second issue raised by provision of economic measures is that of equity. Will offering a subsidy for carpoolers create an issue of equity for non-subsidised employees? Will the inclusion of any travel benefit be problematic in the firm's collective bargaining process? The equity issue must be addressed by the public sector as well. A major policy change was introduced in London by the then Greater London Council under the title "Fares Fair" with sharp reductions in bus and metro fares. It proved difficult to finance and inequitable between different regions of London, and legal intervention led to its ultimate demise. But while it lasted, the programme led to a reduction in cycling which increased again when fares were restored to their original levels. Therefore, the substitution of travel modes was less between car and transit than between cycling and transit.

Third, administrative issues can arise. Employers may incur administrative costs which were not anticipated. For any programme to be successful, full disclosure and understanding of all expected responsibilities and costs is advisable.

Much remains to be learned about the response of travellers to congestion pricing. Further testing of congestion pricing is being conducted under programmes of motorway operators in Europe and the United States. To effectively moderate travel demand during peak periods, charges may have to be set at such high levels that they would be politically unacceptable for fear of causing the traveller to not make the trip at all. Congestion charges have also been criticized for being discriminatory toward the poor, or for placing businesses in the implementing region at a competitive disadvantage. Chances for successful implementation of congestion pricing will be enhanced, if pricing is included as a part of a comprehensive package designed to address regional mobility problems. That package should also include measures to address equity concerns through the use of congestion pricing revenues. Although the technology for electronic toll collection is available, the extension of that technology to more sophisticated applications of congestion pricing has not been tested. Whether the technology is available to accurately assess the right person the right price at the right time is yet to be proved.

There are also a variety of incentives available for public agencies to provide. For example, while most public transport service is subsidized, the ability to directly subsidize users most often involves the reduction in fares for all users or for targeted user groups (e.g. students, elderly, disabled). Fare discounts targeted to commuters are relatively rare because most commuters represent "choice" riders (i.e. have a choice between commute options); commute service is generally the most costly to operate during peak periods; and premium express-type commuter service most often commands a fare surcharge, not a reduction.

One other potential problem with congestion pricing deserves emphasis, that of public perception. Many governments have levied or have considered levying tolls on motorways as a means of financing future maintenance or construction. Confusion could result -- is the programme's objective to price the motorist off the road, or to levy more taxes? A dual objective may be difficult to sell, in particular when proponents of demand-side measures will have a valid point to wait until the effects of tolling can be measured before investing in supply-side improvements.

Examples

Congestion Pricing

Area Licensing Scheme (Singapore) - Introduced in 1975, the objective was to discourage autos from entering the congested central area during the morning peak by requiring purchase of a supplementary license ($2.50 per day) to enter the central area. Following introduction of the ALS, peak hour trips into the central area fell from 56 per cent to 23 per cent of total work trips. By 1983, total vehicle traffic was still 23 per cent of total work trips. In 1989, the ALS was extended to cover the evening peak (see Section III.4 for a more detailed description and review).

A1 (Lille-Paris) Motorway (France) - In April 1992, on the A1 Lille-Paris motorway, a higher toll (25 per cent surcharge over the normal toll) was initiated for traffic driving back to the Paris region during the red period (4:30 p.m.-8:30 p.m.) on Sundays at the toll facility around 60 km north of Paris. A lower toll (with 25 per cent savings) was in place during the green period before and after the red period. An extensive evaluation study has been conducted and shows that some 6 per cent of red period traffic is shifted to the green period.

Parking Pricing

Inner City of Goteborg (Sweden) - The City of Goteborg (population of 450,000) has a policy to reduce automobile in the city centre. A key action to reduce car traffic is the increase in parking fees in order to encourage drivers to avoid parking in the central area. In 1987, the parking fees were increased 100 per cent in the central business district. The immediate effect was a reduction of parking occupancy of 20 per cent. However, after one year the occupancy of parking lots was nearly at the same level as before the increase of the parking fees. In other inner city areas, the increase of the parking fees was up to 400 %. The effect was also an immediate reduction of parkers, and then back to almost the same occupancy.. However, a negative aspect was increase of the traffic on some streets up to 30 %. The reason was more car drivers, seeking for parking lots with lower price.

US WEST, Bellevue (Washington, USA) - This suburban Seattle employer (telephone company with approximately 1,150 on-site employees) charges employees $60 per month (market rate) for parking in its on-site garage. Carpools with 2 persons are charged $45 per month and 3 or more person carpools and vanpools receive free parking. Parking is also constrained; only one space is available for every three employees. The company reserves parking for carpools and vanpools; parking for drive alones, even those who have paid the monthly charge, is on a first-come, first-served basis. The company's drive alone rate is only 26 per cent, well below the 80 per cent rate for the surrounding area.

Hartford Steam Boiler (Hartford, Connecticut, USA) - This enterprise charges employees $110 per month to park, the market rate for spaces the employer leases. Carpools with two persons are charged $75; three persons $40 and four or more charged $10. In addition to the lower fees, ridesharers can split the cost of parking so that each is paying a maximum of $37.50 (2-person carpool) as compared to the drive-alone rate of $110. The company's drive alone rate is only 40 per cent, compared to 63 per cent for a nearby, similar site with free parking.

Copenhagen (Denmark) - In 1990, paid on-street parking was introduced throughout Central Copenhagen (14,000 spaces) with the primary objectives being to reduce commute traffic, to improve traffic flow, and to control parking demand. During the first year of operation, a high rate of non-payers (50 per cent in 1991) necessitated increased parking enforcement, which subsequently reduced the non-payer rate to 13 per cent. The effect of instituting paid public parking has been an up to 5 per cent decrease in traffic demand toward Central Copenhagen. Public transport usage has also increased slightly.

Australian Capital Authority (ACT) (Canberra, Australia) - In 1991, an experiment was begun in ACT offering free parking in car parks in downtown Canberra and at a major regional centre for vehicles with 3 or more occupants. The results were encouraging and consideration is being given to extending the scheme to other centres in ACT.

Section II.6. **Administrative Measures** provides additional examples which specifically address Parking Management Measures.

Transport Allowances

State Farm Insurance (Costa Mesa, California, USA) - State Farm Insurance's Southern California Regional Office is located in a large suburban office park which is highly oriented to auto travel. The building is a one-story structure surrounded by ample, free surface parking for all its 1,000 employees. There is little transit service to the site. Each day, as employees arrive at the company's on-site parking lot, an attendant checks the occupancy of the vehicle and issues coupons worth a particular subsidy amount. The coupons are accumulated by the employee and returned for cash redemption at the close of a pay period. The value of the subsidy increases with the size of the carpool (single-occupant autos receive no coupons). The result of this programme was to increase the average auto occupancy level from 1.22 to 1.55 in two months, a vehicle trip rate reduction of over 30 per cent.

Transit/Ridesharing Financial Incentives

Union Bank (San Diego, California, USA) - Union Bank, located in downtown San Diego, offers its 315 employees a 100 per cent transit subsidy. Employees have free parking, but off-site, in a company-leased garage located several blocks away. Monthly garage pass holders are given passes to the downtown trolley service, which connects the garage to the office. Union Bank has a transit share of 36 per cent, substantially higher than the 19 per cent average for all employers in the CBD. This transit share equates to a 15 per cent trip reduction compared to all downtown employers.

Jet Propulsion Laboratory (Pasadena, California, USA) - The Jet Propulsion Laboratory (JPL) in California spends $250 to begin a vanpool and a $20 per month subsidy to each employee that rides to work in a vanpool. The company supports the programme by fuelling and washing the vehicles, providing preferential vanpool parking and a guaranteed ride home programme. Since the programme's inception in 1989, over 30 vanpools have been formed, increasing the mode share of employees commuting to the JPL in vanpools from one percent to six percent. The employees share of the vanpool fare ranges from $60 to $80 per month, depending on the commuting distance travelled.

Public Transport Pass Programmes

Carte Orange Travel Pass (Ile-de-France, France) - The Carte Orange travel pass was initiated in the Paris metropolitan region in 1975 and allows travellers to use all public transport services (e.g. metro, buses, regional trains) of different authorities in the region during a one-week or one-month period with a special coupon, valid for the pre-selected zones. As mentioned above, since 1982, employers located in the Paris transport region have also contributed to the Carte Orange Travel Pass purchase price by employees.

<u>Job-Tickets in Germany</u> - In many German cities, some private and public employers have contracts with public transport companies which accept the firm card as a transit pass. The employer pays 50 to 90 per cent of the least expensive normal pass per employee. Thus, the public transport company has a guaranteed income every month, the employer can reduce parking facilities, the employee has a free transit pass, and the modal split improves which is beneficial for the general public.

<u>Fare Share Programme (Montgomery County, Maryland, USA)</u> - Montgomery County provides discount transit passes through employers, to people who work in the County. Employers participate in administering as well as share in funding the discount passes. The County and the employer each provides a 25 per cent subsidy. Employers purchase the discount passes directly from the County and employees receive a 50 per cent fare discount, and can purchase fare media directly from the employer; a convenience feature. Employers determine how many passes to purchase, pay for passes in advance of selling them to employees, and handle cash transactions associated with selling fare media to employees. Employers also report the number of participants to the County.

<u>Corporate Transit Pass Program (Los Angeles, USA)</u> - The Southern California Rapid Transit District initiated a service-areawide employer pass programme in which over 700 employers, representing 50,000 employees now participate. Approximately 500 employers subsidize the passes, on average by $32.40. SCRTD supports the pass programme by assisting employers to develop travel demand management plans, providing individual routing assistance to employees, and participating in employers' transport information efforts.

Innovative Financing

<u>Versement transport (transport tax) (France)</u> - As mentioned above, since 1971 in the Paris metropolitan region and since 1973 throughout the remainder of France, authorities have set up a system whereby public transport can be partially financed by employers (by law, only those with more than 9 employees). All urban areas with a population over 100,000 and most of those between 30,000 and 100,000 are currently taking advantage of "versement transport." At the current time, the maximum rates are between 0.5 and 2.2 per cent of each employee's salary (paid for by the employer).

Selected references

1. SETRA (1993). *Evaluation quantitative d'une expérience de modulation de péage sur l'autoroute A1*. Etude réalisée par S.E.E.E. Infra pour le Service d'Etudes Techniques des Routes et Autoroutes. Bagneux.

2. DREIF (1992). *Les transports de voyageurs en Ile-de-France 1991*. Direction Régionale de l'Equipement de l'Ile-de-France. Paris.

3. U.S. DEPARTMENT OF TRANSPORTATION (1992). *Suburban Parking Economics and Policy: Case Studies of Office Worksites in Southern California*. Washington D.C.

II.6. ADMINISTRATIVE MEASURES

Description - Administrative measures are defined as organisational and legal agreements and structures that underlie or help to enhance implementation of congestion management measures in a local area. They are not themselves pure transport demand management or transport system management measures, but rather are partnerships, regulations, and other institutional, organisational, and legal structures to promote, implement, and monitor congestion management measures. The administrative class of measures includes five basic categories:

♦ Transportation partnerships, i.e. employer-formed groups aimed at addressing collectively local transport/growth management issues (such as congestion) and at assisting employers and developers to comply with transport regulations or ordinances. Two common names for these partnerships are transportation management associations (TMA) and business roundtables.

Parking guidance system (Aachen, Germany)

- Trip reduction ordinance, a legal mechanism that requires employers (and sometimes developers and property owners) to implement trip reduction programmes at individual worksites, or employment, residential, and commercial developments.

- Alternative (or flexible) work schedules, including three non-traditional work schedules: (1) flextime, in which employees pre-select their work-hours within time constraints imposed by their employer so that they arrive at the worksite earlier or later than is traditional which helps to reduce congestion during peak commuting periods); (2) compressed work weeks, in which employees work longer, but fewer, days per week (reducing the number of commute trips made); and (3) staggered work shifts, in which employers stage the time of arrival of various groups of employees (to reduce arrival peaking). A fourth type of work schedule involves telecommuting, in which employees work at home or in a work centre nearer their homes (and therefore reduces the number of trips and/or the distance travelled). Telecommuting is covered in Section II.3. **Telecommunications Substitutes.**

- Auto-restricted zones (ARZ), i.e. designated areas in which vehicular traffic is mainly prohibited or severely restricted. In some cases, private motorized vehicles are prohibited from the area but with accommodations for emergency access. In most applications, the speed or nature of traffic in an area is constrained (e.g. Woonerfs and other "traffic calming" arrangements).

- Parking management measures targeted at controlling the supply and usage of the finite supply of parking within an urban area (see also Section II.5. **Economic Measures** for a discussion of parking pricing).

Objectives and Major Impacts - The primary objective of administrative measures is to facilitate the effective implementation of mainly peak period car trip reduction measures. This is done in several ways: by requiring employers and other groups that have the opportunity to influence behaviour to implement demand management measures; by forming partnerships of employers and other interested parties to increase the efficiency and effectiveness of demand management implementation; by instituting flexible working hours and worksite services to allow employees to use commute alternatives more easily; and by restricting automobiles from designated areas, thereby enabling freer movement of the preferred mode of travel (e.g. public transport, bicycle, walking).

Application of Measure - In general, administrative measures are targeted to commuter travel except perhaps ARZs and parking management measures which target the convenience of shoppers and residents as well.

Transport partnerships and associations typically serve the following functions:

- Information exchange among members,

- Assistance to individual member employers with development of travel demand management (TDM) plans, regulatory compliance, and solutions to site-specific problems,

- Promotion or provision of TDM services and incentives,

- Advocacy for improvements in transport service and facilities.

They can also include coordination of travel demand management support programmes. An example would be a guaranteed ride home (GRH) programme in which free or subsidized transport is offered to employees who use a commute alternative but who need to leave work during the workday due to a personal emergency, such as illness, or who must work overtime. Transport typically is provided by taxi, rental car, or company security personnel.

Trip reduction ordinances/regulations typically apply to employers over a certain size or to promoters of development projects over a certain area/space (in the Los Angeles region, the regulation applies to employers with at least 100 employees at a single worksite). The ordinance usually specifies actions employers must implement, sets a target trip reduction to be met, or simply requires employers to develop trip reduction plans (in Los Angeles, the objective is an Average Vehicle Ridership for employees of 1.5).

The applicability of alternative work schedules is directly related to each individual employer's situation. Compressed work weeks are most appropriate for office and administrative functions and for line and piece manufacturing processes. Flexitime is most appropriate for offices and among administrative and information workers. Staggered hours are most appropriate for offices and piece manufacturing.

Auto-restricted zones (ARZ) have been applied in many types of land use situations ranging from residential and commercial/ historic districts to institutional areas. Numerous Italian cities (e.g. Milan, Bologna) and Swiss cities (e.g., Berne, Zurich, St. Gallen, Geneva) have introduced an ARZ with access permitted only for pedestrians, bicycles, taxis, and public transport vehicles. In conjunction with the ARZs, additional parking facilities have sometimes had to be constructed at the border of the ARZ.

In several European countries, most notably Germany and France, "Zones 30" were established (i.e. areas where the speed limit is 30 km/h). The primary reason for the limited speed zones was safety (for both the motorist and the pedestrian). The Netherlands, which pioneered the concept of "woonerfs" in the 1970's also has numerous residential areas with "calmed" traffic.

Institutional Responsibility for Implementation - Institutional responsibility for administrative measures varies by measure. Responsibility for implementing alternative work hours and support programmes frequently rests with employers, although developers and property managers also may be involved in providing on-site services.

Transport partnerships, associations, and city revitalization groups (e.g., chambers of commerce) are generally initiated by employers but may be started also by developers, especially in new or developing areas (e.g. Warner Centre Transportation Management Organization, California USA). Government agencies sometimes encourage or foster employer associations' development through seed funding programmes (e.g. Seattle METRO, USA).

Trip reduction ordinances/regulations are most often initiated by cities, counties, air pollution control districts, and other local government entities. Many recent ordinances have been developed in order to comply with environmental initiatives.

Alternative work schedules are always implemented by employers at worksites.

Auto-restricted zones (ARZ) are most often implemented by a regional or local public agency with authority over the operation of local streets. The successful ARZ applications have almost invariably involved a local business association, some commitment of funds from the private sector, and the formation of an ongoing public/private task force. In some cases, a special funding source was created.

Effects on Travel Patterns - The impact of most administrative measures is assessed through the effectiveness of economic actions taken, information services, or other activities which the administrative measures seek to complement. For example, the mere existence of a Transportation Management Association (TMA) or trip reduction ordinance has little impact on travellers' mode choice. If, however, the administrative measure initiates or contributes to initiate an effective trip reduction action that would not otherwise have occurred, the administrative measure is indirectly responsible for the change in the travel pattern. For example, if trips are reduced because a TMA subsidizes transit passes or a trip reduction ordinance requires employers to restrict parking at a worksite, these administrative measures can be termed successful, even though it was actually the subsidy or parking restriction that influenced the traveller to change modes.

A growing body of research suggests that participation of the business community, which has some influence on commuters' mode choice, is important to successful demand management implementation. Thus, ordinances that require employers to implement demand management measures should be effective in reducing trips.

Alternative work hours programmes, if implemented on a wide-scale, can have a significant and quite direct impact on vehicle trips. A comprehensive evaluation study for Ottawa, Canada, where approximately one-half of the downtown work force had staggered or flexible work hours, found the following:

- ◆ peak-hour worker arrivals and departures declined about 20 per cent,

- ◆ peak-hour bus loads and vehicle parking volumes declined nearly 20 per cent,

- ◆ peak-hour traffic volumes at the CBD cordon declined 10 per cent,

- ◆ average travel times declined almost 12 per cent during the peak hour (and six per cent overall for the two-and-one-half hour peak period).

An auto maker in Utsunomiya City, Japan introduced flex-time and enjoyed great results. For example, commute times by bus dropped from an average of 47 minutes to 28 minutes. It should be noted that with alternative work hour arrangements such as compressed work weeks, other non-work (e.g. leisure) trips could be created.

With the institution of parking management along two major axes in Paris, 27 km long (termed the Axes Rouges programme), average travel speed increased by 14 per cent on Axes Rouges and 62 per cent of the affected motorists declared that traffic flow was better.

Cost-Effectiveness - The costs to implement a trip reduction ordinance fall on several groups: the implementing agency, which bears the cost for administration; employers, who bear the cost of complying with the regulation; and in some cases, support groups, such as public transport agencies, which must provide additional services to assist employers to comply with the regulation. Costs vary substantially, depending on the number of employers (and employees) affected, the level of support offered to employers in complying with the regulation, and the specific provisions of the regulation.

In U.S. cities which administer their own travel demand management ordinance, costs range as high as $2.60 per employee affected. Budgets for other agencies administering programmes range widely around this figure. Some programmes completely recoup administration expenses through employer fees, but most recover no more than a small percentage of their costs.

There is little documentation on the costs for implementing alternative work hour programmes. Basic cost items will include labour time to plan and set up the programme, and possible increased utility and security costs associated with opening an office earlier and keeping it open later than usual. The principal drawback to alternative work schedules is the common perception by management that face-to-face contact is a necessary component of an individual's job, no matter what the occupation. Indeed, a study of firms in the downtown of a major metropolitan area in the U.S. found that 15 per cent of the employers involved in staggered work hours reported workday communications problems; however, they also stated that these costs were balanced by other efficiency gains in increased employee punctuality.

Special Problems or Issues - The most important issue to be remembered with administrative measures is that they themselves are not trip reduction measures. They can be effective only if they encourage effective actions that would not otherwise have occurred. This is especially true for trip reduction ordinances, which must be written to require performance (i.e., meeting a trip reduction target), not merely a process (developing a trip reduction plan). Similarly, TMAs and other partnerships will be successful only if they have the financial and political capital/support to implement effective travel demand management measures or to influence implementation of these measures by other groups that have the power to influence travel behaviour.

There are particular concerns with ARZs which must be addressed. Prior to implementation of any ARZ project, economic impacts must be investigated. For example, an additional parking fee of 25 SEK per day in the centre of Stockholm was forecast to lead to an 8 per cent reduction in retail trade customers, certainly a significant drop.

Examples

Transportation Partnerships and Associations

Warner Centre Transportation Management Organization (TMO) (Woodland Hills, California, USA) - The Warner Centre TMO, which serves a suburban mixed use centre within the Los Angeles metropolitan area, actively promotes commute alternatives through worksites and assists employers to develop and implement TDM programmes. The TMO also offers direct commute benefits: multi-company vanpool programme with substantial financial incentives, subsidized express bus service from several residential areas, ridematching for employees of member companies, bicycle club, and other services to employers and employees. Over 80 per cent of large employers (more than 100 employees) participate in the TMO's programmes.

Seattle Metro, Easy Ride (King County, Washington, USA) - Seattle Metro implemented a Guaranteed Ride Home (GRH) programme in suburban areas in King County. To be eligible, commuters are required to rideshare at least three days per week and register with Metro for the programme. Transportation is provided through via taxi and commuters are reimbursed through a voucher system. Commuters can use up to 40 miles of taxi travel, about 4 average trips. Participants have indicated the programme is important in their decision to rideshare; 69 per cent of the participants, including commuters who shifted from the drive-alone mode to high-occupancy vehicles when they joined the programme, indicate that the GRH programme is somewhat or very important in their decision to continue to take the bus, carpool, or vanpool to work. 22 per cent rated it very important.

Trip Reduction Ordinances/Regulations

<u>South Coast Air Quality Management District (SCAQMD), (Los Angeles area, USA)</u> - The SCAQMD adopted Regulation XV in late 1987. The regulation, covers a four-county district of Los Angeles, Orange, San Bernardino, and Riverside counties and affects approximately 9,000 worksites. Employers with 100 or more employees at any one worksite are subject to the regulation, which requires employers to develop and implement trip reduction plans to increase average vehicle ridership (AVR) at the site to 1.5 persons per vehicle. According to a 1992 study by Ernst & Young, area-wide AVR has increased from 1.1 in 1988 to 1.31 in 1992, an average increase of 4.6 per cent per year.

Alternative Work Schedules

<u>Ministry of Transport and Public Works (The Netherlands)</u> - The Ministry permitted all employees to switch from a traditional, five-day 38-hour work week to a schedule with four 9-hour working days per week (with either a proportional income loss or the need to work two extra days per month). The programme did not produce any changes in the reported commute mode split for the workers but did result in a 10 per cent reduction in commuting traffic. The potential for increased travel on the extra day off was not evaluated.

<u>Downtown Ottawa (Canada)</u> - Approximately one-half of the downtown work force has staggered or flexible work hours. Their effects are wide-ranging (See above description).

<u>Flexible Work Hours in Switzerland</u> - Many Swiss companies and almost all public sector agencies have introduced flexible or freely arrangeable work hours. The obligatory block-hours are between 8:30 a.m. and 11:15 a.m. and between 2 p.m. and 4 p.m. The employees are free to work any hours between 6:30 a.m. and 6:30 p.m. for a total of 42 hours per week. The results have been positive in terms of reducing peak-period congestion.

<u>Town of Onteniente (Province of Valencia, Spain)</u> - Valencia has instituted staggered work hours within an industrial zone (primarily textile factories) affecting more than 1,000 workers. The project was initiated when a bridge connecting the two parts of the town was closed in 1989 for repairs.

Auto-Restricted Zones

<u>Inner City of Visby (Sweden)</u> - The town of Visby was a major commercial centre in the Middle Ages and a ring wall, constructed in the 13th century, is still preserved. The population is 22,000 but during the summer months, the influx of visitors causes traffic levels to double. Since 1987 motor traffic to or from the inner city has been banned for "non-legitimate" traffic for a period of nine weeks (around mid-June through mid-August). Legitimate traffic is considered as transport to own residences in the inner city and to work places (if a reserved parking facility is provided), as transport of handicapped individuals, for service and maintenance, and for public transport (buses, taxis, etc.). Parking lots for around 1,000 vehicles are available outside the ring wall. The restrictions are announced by signs at the gates. Motor traffic has been reduced more than 40 per cent. However, a survey showed that 30 per cent of the private sector motor vehicles could still be considered non-legitimate. Similar approaches have been tried by other European cities with historic centres which generate tourist traffic.

<u>Closure of Inner-City Centres in Aachen and Lubeck (Germany)</u> - These two cities have been the first to close the inner city centre for private vehicles on Saturdays. In Aachen, for example, the restriction lasts until 5.30 pm on one Saturday per month and until 2.30 pm on the other Saturdays. The inhabitants are the exception and are permitted to drive within the city centre. Inhabitants and shoppers were strong proponents from the outset. Merchants and traders were opposed at the beginning, but have become more accepting as the ARZ benefits have materialized. Auto usage for visitors has dropped 8 per cent. Noise levels have decreased and air pollutant emissions have been reduced.

Parking Management Measures

Axes Rouges in Paris (France) - Axes Rouges consist of the prohibition of all parking and stopping along designated main roadways. Deliveries are displaced to side streets. The first 27 km were instituted in 1990 along two axes in Paris. (See above for description of results).

Signal Park (Munich, Germany) - A parking guidance system named "Signal Park" has been recently introduced at Munich's new airport. The system monitors more than 9,000 parking spaces and guides drivers by the quickest route to the nearest unoccupied parking stall. The system is estimated to save each parking motorist an average of three minutes in locating a parking stall.

Parking Management in Enschede (The Netherlands) - By means of restricting usage of public parking spaces, instituting higher parking fees, and improving enforcement, Enschede achieved roughly a 50 per cent reduction in the number of inner-city visitors arriving by automobile. The use of public transport increased from an average of 13 per cent to 13.5 per cent. Several other Dutch towns follow similar techniques, making the approach to the town centre extremely difficult or impractical for visiting motorists.

On-Street Parking Enforcement (London, UK) - As part of its "Red Route" Pilot Project, the Traffic Director of London identified locations where existing on-street parking at important locations created significant traffic congestion problems, especially during peak periods. At one commercial area near an intersection, long vehicle queues were created by allowing store owners to park their vehicles without adequate restrictions and enforcement. Working with the local community traffic officials, the Traffic Director of London placed an all-day no-parking restriction at this location supported by strong police enforcement and costly fines for violators. As a result of this simple action, an additional lane of capacity was restored at the intersection enabling traffic flows to improve significantly, delayed only slightly by the signalized intersection. The additional capacity also enabled the Traffic Director, working with officials from the London Transport, to install a curb-side lane for exclusive bus use only during the day. In this case, removal of on-street parking supported by strong enforcement was considered to be a simple yet highly effective measure to reduce congestion and improve the flow of autos and buses (Section II.9. **Public Transport Operations** describes public transport service improvements also made along the Red Routes in London).

Parking Management Measures (Berne, Switzerland) - One primary objective of Berne's transportation policy is to promote "traffic that is adapted to the city rather than to adapt the city to the traffic." Consequently, the City of Berne was the first Swiss city to introduce a system favouring local residents in the use of public parking lots. The residents are also eligible for a parking permit which entitles them to unlimited parking in the so-called "blue zones" which normally restrict parking to a 1.5 hour limit. This system not only keeps commuters out of the urban residential districts, but also improves the parking possibilities in these areas for residents and visitors. It also helps preserve the attractiveness of the old city as a service and shopping centre well served by a dense and attractive existing public transport system. The policy also resulted in the number of parking spaces in the city area being consequently reduced, parking fees as much as tripled and in many spaces being limited to only very short-term parking (i.e. less than 30 minute). The result of the overall parking policies has been a dramatic reduction in motorized traffic in the city centre since the 1960's when the City of Berne was choked with private cars.

Residents-Only Parking Areas (Germany) - Besides other parking restrictions like increased parking pricing, an increasing number of German cities divide their city centres into different sectors where only residents of the area are permitted to park their car. The cars are marked with a badge which costs for example $50 per year. Visitors, commuters, and shoppers have to park in designated parking lots or in on-street parking stalls with parking meters.

1. U.S. DEPARTMENT OF TRANSPORTATION (1992). *Transportation Management Associations in the United States*. Washington D.C.

2. U.S. DEPARTMENT OF TRANSPORTATION (1984). *Implementation of Downtown Auto-Restricted Projects*. Washington D.C.

II.7. ROAD TRAFFIC OPERATIONS

Description - The motorway/road operations strategies encompass all measures which manipulate the road system to encourage, enable, or force the most efficient traffic. The result should be optimised traffic flow, i.e. with the least amount of delay. Such measures have been applied for years either sporadically or not systematically, and as a result with relatively localised or temporary effects. They include:

♦   Entrance Ramp Controls (or Ramp Meters): These are traffic signals placed on freeway/motorway entrance ramps that supply traffic to the motorway in a measured or regulated amount. Meters can be operated to discharge traffic at a rate so as not to exceed a downstream bottleneck. By not exceeding capacity, throughput is maximised, speeds remain uniform, and congestion-related accidents are reduced.

♦   Traveller Information Systems: They provide a variety of information that assists travellers in reaching a desired destination via private automobile, public transport, or some combination. Some systems can provide information to motorists regarding their location (via a navigational system) and continuous advice regarding traffic conditions, alternate routes, weather conditions, warnings, and parking conditions (note that Section II.4. **Traveller Information Services** focused on pre-trip information dissemination; this section addresses systems which provide information while the motorist is enroute).

♦   Traffic Signalisation: Traffic signal system improvements have three fundamental elements. The first involves the coordination of groups of signals by using either interconnection or highly accurate time-based coordination. The second involves systematically optimising the signal timing parameters. The third involves the use of advanced traffic control functions by using a master computer control.

♦   Motorway Traffic Management Systems: The objective of such system is to balance the traffic demand between motorways and the adjacent arterial street system by means of both traffic control technologies (i.e. entrance ramp controls, mainline controls) and driver information systems. The basic concept is to predict demand and monitor operating conditions on the overall system and to transmit real-time information on congestion and/or institute control measures as appropriate. Its dynamic traffic control system responds to changing traffic conditions across different jurisdictions and types of roads by routing drivers around delays where possible. A motorway traffic management system integrates management of various roadway functions, including ramp metering, arterial signal control, speed control, and traffic responsive route guidance (e.g. variable message signs). The concept is in common practice as part of tunnel and viaduct control systems.

♦ Incident Management: This includes the spectrum of activities involved in detecting, responding to and clearing roadway incidents. It is the coordinated and preplanned use of human and technological resources to restore full roadway capacity after an incident occurs, and to provide motorists with information and direction until the incident is cleared. The duration of the incident is minimised by detecting and responding to the incident as soon as possible, and the expeditious clearance of any obstructions from the travel lanes. Restoration of full capacity is achieved by removing vehicles both from the travelway and from the sight of the motorist. Capacity restoration must also await full dissipation of the queue caused by the incident.

For a variety of reasons -- including existing technological constraints and the current limited resources devoted to incident management -- there is only limited information being provided to motorists to date regarding route diversion possibilities. The best current practices inform the motorist about the distance to the incident site and about the estimated delays. Many efforts are currently underway to improve this information dissemination. An equally important element of a region's efforts to manage traffic congestion should be the dedication of sufficient resources to undertake a comprehensive and effective safety improvement programme and thereby reduce the number of incidents.

♦ Traffic Management at Construction Site: Concurrent with any major construction effort, steps must be taken to effectively manage the affected travel demand (with due consideration given to safety, mobility, neighbourhood impacts, business impacts, goods movement, etc.). The programme should include a composite of the other demand management and systems management measures described throughout this chapter. The programme should also try to minimise the time needed for construction and hence diminish traffic delays at the construction zone. Various techniques have been successful in reducing the time required to complete a project, including: (1) working at night or on weekends; (2) staging work to minimise shifts in traffic patterns; (3) providing incentives for the contractor to complete the project early and penalties for late completion; and (4) using new construction/maintenance techniques and materials (e.g. faster-setting concrete).

♦ Reversible Lane Techniques: A simple, flexible approach to managing unbalanced traffic demands during the peak hours is to provide a reversible lane which operates in the predominant direction during the peak periods. The technique can be applied on a regular (i.e. daily) basis or can be adapted to accommodate special events. The primary limitations to the application of reversible lanes are threefold. The directional distribution of the traffic demand must be sufficiently unbalanced to enable acceptable operations in the "off-peak" direction. Second, implementation of the techniques requires the presence of on-site personnel or the development of the capability to use remote control of the necessary traffic control devices. Third, concerns over safety must be addressed and overcome; for example, any confusion on the part of the motorist could contribute to unsafe manoeuvres.

Objectives and Major Impacts - The basic objective of both motorway and surface street traffic management measures which are geared to address recurring congestion problem areas is to expand the operating capacity at these bottleneck locations.

The objective of entrance ramp controls is to significantly improve mainline traffic flow, in particular in terms of increased travel speeds and improved travel time reliability. The effects of entrance ramp controls along the Boulevard Périphérique of Paris were a 15-18 per cent increase in mainline travel speeds.

Evaluation of the effectiveness of traffic signal system improvements has been undertaken in numerous locations. For example in France (where the SURF, SAGE, and GERTRUDE systems have been in place for some time), the optimisation of fixed-time signals yields a 25 per cent improvement in speed when compared to an uncoordinated condition.

The effects of large-scale motorway traffic management systems have been well-documented. Based on operating systems, the average motorway mainline speeds typically improve as much as 30 per cent. After accounting for the additional delays absorbed at the entrance ramp meters, the average overall speed generally increases roughly 20 per cent. Note that these effects apply only to the peak directions of traffic flow during the peak periods.

The effects of traveller information systems are difficult to quantify but its potential to have significant impacts is evidenced by a survey of French motorists who indicated they changed their travel route because of variable message sign displays at least once every two weeks. In Germany, a system of variable message signs for speed control and congestion warning is credited with reducing traffic accidents on the A5 motorway by nearly 50 per cent.

The extensive incident management programme in the Los Angeles metropolitan area has been estimated to save 65 per cent of the travel delay otherwise caused by non-recurring congestion. The Chicago Area Expressway programme achieved a 30 per cent reduction in peak period congestion.

Institutional Responsibility for Implementation - For motorway operations measures, the implementing agency is the operating agency responsible for the motorway system. In addition, the successful programmes have involved (throughout the development and implementation process) all officials and agencies with a vested interest, including: elected officials; highway operating agency;

Reversible roadway lanes for additional capacity (Barcelona, Spain)

public transport operators; local jurisdictions; police departments; independent authorities (i.e., those with responsibility for the operation of toll roadways, bridges, or tunnels); fire and rescue; emergency services; environmental protection agencies; and towing services. For some measures to be effective, they must also incorporate the concerns of other constituencies, including: neighbourhood and civic groups, chambers of commerce and business organisations; emergency vehicle operators (e.g., police, fire, ambulance); utility companies; automobile clubs; major trucking companies and associations; school officials; officials responsible for the designation of safe routes for the transport of hazardous materials; and the media.

Cost-Effectiveness - Motorway/road operations measures are cost-effective. Experience in major metropolitan areas indicates a return of $17 in benefits for each $1 invested in a motorway surveillance and incident management programme. However, most of these technologies, although functional and available since the 1970's, have not reached their full potential due to their perceived high implementation and operating costs and to only sporadic and non-systematic applications.

An effective incident management programme can be tailored to involve the acquisition, installation, and operation of off-the-shelf equipment. Recent advances in technology can improve the effectiveness of the incident management programme, but the initial, major benefits are achieved through coordinated, cooperative, and systematic approaches taken by the various jurisdictions involved.

Special Problems or Issues - A major issue that is often raised in conjunction with ramp metering is the potential for an undesirable number of motorway trips to be diverted to the adjacent surface street system. Experience clearly demonstrates, however, that a well designed and operated ramp metering system improves overall operations of the motorway system and does not cause significant diversion to parallel streets. In many applications, the motorway mainline congestion was sufficiently relieved by ramp metering and ramps eventually carried increased levels of traffic volume after institution of the metering.

The main potential problem with traveller information systems is the type of information to be provided to the motorist and the accuracy of such information which is a direct function of the updating provisions of the system. In addition, as traffic congestion problems worsen, potential solutions can be as numerous as the number of affected motorists.

Traffic control system improvements are less hampered by institutional impediments than many other types of supply management actions. Even so, the quantity and quality of traffic control system programmes are impeded by such factors as interjuridictional coordination problems, the lack of traffic engineering staff skills in some jurisdictions, and the lack of priority given to improved traffic control programmes.

The greatest benefits in an incident management programme (next to reducing the number of incidents) are achieved by reducing the duration of the incident. One approach which has been successfully taken is to remove the disabled vehicle from the travelway as quickly as possible, even to the possible physical detriment to the vehicle or its contents. For both commercial and private vehicles, the cost of the excess damage, even though it is minor when compared to the "public good," is nevertheless an unwelcome expense to an individual, business, or insurance company. The question of liability associated with fast removal policies must be addressed and resolved.

Entrance Ramp Controls

Ramp Metering (The Netherlands) - This has been installed in Delft and Amsterdam. The Delft application increased motorway-capacity by 5 per cent and an overall reduction in motorway delays. Enforcement was a problem; 15 per cent of the traffic on the ramps ignores the traffic signal. The Amsterdam application reported that motorway capacity has not been improved but average travel speeds increased dramatically. There was some diversion of traffic away from the corridor. Violations are not as serious as in Delft (less than 5 per cent of the motorists pass through the red traffic signal indication).

Boulevard Périphérique in Paris (France) - Different entrance ramp control algorithms were tested on the ring road. The selected ramp metering resulted in nearly a 16 per cent reduction in overall motorway travel time and the total time spent in congestion decreased over 50 per cent. The analysis also found that the quantifiable benefits are twice the investment and operations costs of the system.

Copenhagen Ramp Metering (Denmark) - Ramp metering was installed along an inbound route in June 1992. The motorway has a 4-lane cross-section and peak traffic flow rates of 4,300 vehicles per hour downstream of the ramp metering. The ramp itself carries 800 vehicles per hour. Initial observations are that the motorway volume has slightly increased during the peak hour. At the same time, the ramp volume decreased slightly. Motorway speed has increased from about 50 km/h prior to ramp metering to 75-80 km/h after implementation.

Detroit Ramp Metering (USA) - The Michigan Department of Transportation Surveillance Control and Driver Information (SCANDI) ramp metering began in November 1982 with six ramps. Another 22 ramps have been added to the system since then. A research study determined that ramp metering increased mainline speeds by about 8 per cent. Peak hour volumes on the mainline reached 6,400 vehicles per hour in a 3-lane section. The total number of accidents has declined nearly 50 per cent and injury accidents are down 71 per cent.

San Diego Ramp Metering (USA) - Ramp metering was initiated in 1968 and now includes 81 metered ramps. Mainline volumes have stabilized at 2,200-2,400 vehicles per hour per lane. A noteworthy aspect of the programme is the metering of four freeway-to-freeway connector ramps. Metering freeway-to-freeway connectors requires careful attention to storage space, advance warning, and sight distance.

Traveller Information Systems

Route Information Amsterdam (The Netherlands) -- Route Information Amsterdam (RIA) is the first application of variable message sign technology for dynamic route information in the Netherlands. Motorists are provided information regarding traffic congestion (e.g. queue length, travel time and speed) prior to two tunnels which constrain traffic flow.

POINTER Project (Japan) - The POINTER Project (Positioning and Orienting Information for Traffic En Route) was established to create an easy-to-understand system of road information, linking route numbers, signs, kilo posts, and maps. The system was also designed to be compatible with the eventual widespread implementation of advanced transport technologies and systems.

Advisory Radio (Germany) - Advisory radio messages are broadcast at regular intervals concerning congested areas along the dense Autobahn network. They provide information on the location and length of the congestion and suggestions for alternative routes. Anticipated improvements to the system will be automatic congestion measurements and information dissemination by radio-data-systems with traffic message channel (RDS-TMC). Also, radio frequencies are reserved in the vicinity of major tunnels in several European countries.

SIRIUS System (Ile-de-France, France) - SIRIUS is a traffic operations system on main expressways of the Paris metropolitan area (Ile-de-France) including traffic measurements (loops), surveillance (cameras), control (ramp metering), and information (variable message signs). It is aimed at reducing congestion by modifying travel routes, at providing users with the comfort of real time information, and at improving safety by warning about downstream accidents and congestion.

STORM (Stuttgart Transport Operation by Regional Management) (Germany) - The STORM programme will provide assistance to the traveller when selecting the mode of transport, the time to make the trip, and the route to take. Six pilot projects are currently planned. These include (1) a travel information system for pre-trip planning; (2) a route guidance system for en-route planning; (3) a dynamic P&R (public transport) information system; (4) a public transport connection information system for passengers waiting at stops or transfer points; (5) a fleet management system for goods movement vehicles; and (6) an emergency call system for incident detection.

Boulevard Périphérique Corridor (Paris, France) - Information is given to drivers about congestion and roadway construction activities on the Boulevard Périphérique. A total of 29 variable message signs (VMS) have been installed above the Boulevard Périphérique roadway, 99 VMS on access lanes, 103 VMS on parallel urban boulevards for information, and 50 for directional guidance. By the end of 1994, the total number of VMS will be 350 (versus 281 in 1993).

Traffic Signalisation

Weather-Dependent Traffic Signal Timing (Copenhagen, Denmark) - The computerised traffic signal timing patterns in Copenhagen were found to not be responsive to motorist needs during inclement weather, such as heavy rain or snow, because traffic speeds tend to decrease during bad weather. When average traffic speeds are observed to decrease at least 10 per cent, the signal cycle time and phasing are modified accordingly. As a direct result of this programme, queue lengths, delay, and number of stops on arterials have been reduced during bad weather conditions.

Urban Signal Control Systems (France) - In Paris, SURF (a fixed-time control system with microcontrol possibilities) selects the proper signal timings and phasing in response to detected traffic volumes at 530 intersections. As direct result of SURF, mean vehicle speeds in the controlled area have increased from 17 km/h to 18.6 km/h (a 10 per cent improvement). SURF is now completed by the SAGE expert system to deal with congestion. About 50 systems of the SURF type are in operation in France. GERTRUDE has been implemented in Bordeaux and Reims (and in Lisbonne); it was tested most recently in Paris. The pre-fixed settings are actuated according to measured traffic information with the added feature to provide assurance that intersection blocking does not occur. The PRODYN real time control system was tested in Toulouse and is now operating in Niort.

Motorway Traffic Management Systems

M7 Motorway Congestion Warning System (Budapest, Hungary) - The system was installed in 1988 along a critical upgrade along the M7 Motorway between Budapest and Lake Balaton. The upgrade is 3,500 meters in length, with an average ascent of 3.2 per cent (with a maximum of 4 per cent for 800 meters). About 11 per cent of all accidents occur during Sundays and holidays and along the overloaded "passing" lane. The primary effects were improved traffic safety, reduced congestion, and diminished noise levels. The success of the M7 system has led to the planned installation of an automated traffic management system along the M0 Ring bypassing Budapest (part of the E60 European Highway).

Changeable Direction Signal Systems (Germany) - There are over ten changeable direction signalling systems for parallel autobahns in operation or planned for implementation by 1995. The systems suggest alternative routes using variable message signs on the basis of detected traffic flows on the alternative routes. In existing systems, diversions of 25 per cent have been measured.

Variable message sign before entering Boulevard Peripherique (Paris, France)

INFORM, Long Island (New York, USA) - The Information for Motorists (INFORM) system is designed to improve motorist travel times along the 35-mile central corridor of Long Island. It is the only operating system in the United States integrating the surveillance and control of three freeways (Long Island Expressway, Northern Parkway, and Jericho Turnpike) with adjacent cross and selected parallel arterial streets, to facilitate corridor traffic flow. The freeway and surface street systems are integrated at a Traffic Information Centre in Hauppague, NY.

SMART Corridor (Los Angeles, USA) - The SMART Corridor is a joint demonstration project located along 12.3 miles of the Santa Monica freeway corridor in Los Angeles. The objectives of the SMART Corridor are to provide congestion relief, reduce accidents, reduce fuel consumption, and improve air quality. This will be accomplished using advanced technologies to advise travellers of current conditions and alternate routes (using systems such as Highway Advisory Radio (HAR), Changeable Message Signs (CMS), kiosks, and teletext), improving emergency response, and providing coordinated interagency traffic management. (The freeway system will be operated by the State and the surface streets by the City, with coordination provided via voice communication and electronic data sharing).

Incident Management

Incident Management Systems in Australia - Automatic Incident Detection (AID) systems are installed on motorways in Melbourne, Sydney, and Brisbane. In Melbourne, incident management utilizes not just the AID system, but also the real-time signal coordination system (SCATS), an incident management team, and incident response teams; incident management and coordination of all support services is the responsibility of VIC ROADS (the Victoria State Road and Traffic Authority).

Los Angeles Area Freeway Surveillance and Control Project (Los Angeles, USA) - The project includes police personnel and highway maintenance personnel to deal with an incident, including the management of traffic around the incident. It includes a traffic operations centre for the region's highway system, which collects information from loop detectors, traffic meters, roadside callboxes, and closed-circuit television cameras. The programme is known as CLEAR (Clear Lanes Efficiently and Rapidly) and has over 3,500 traffic diversion plans (i.e. contingency plans). Information is transmitted to motorists via standard variable message signs mounted on pick-up trucks as well as highway advisory radio (HAR).

Chicago Area Expressway Program (USA) - This programme covers 80 miles of freeway and has an annual operating budget of $3.5 million funded from state motor fuel taxes. The programme has 35 heavy-duty tow trucks, several large recovery trucks, cranes, and other specialised equipment.

Traffic Management During Major Highway Reconstruction

Texas Contractor Incentive Program (USA) - A total of 58 incentive/disincentive contracts were reviewed by Texas Transportation Institute. The experience to date indicates that such projects can be completed in approximately half the time of a typical contract. However, it is generally conceded that the final cost for construction will be 10 to 20 per cent higher under an incentive/disincentive contract. Texas uses this incentive programme only for those projects whose construction would severely disrupt traffic or transport service, significantly increase user costs, create safety problems, substantially affect business, or whose early completion would provide a major improvement in the region's transport system.

Metropolitan Transportation Authority Green Line (Los Angeles, USA) - A segment of the heavy rail transit system being built in Los Angeles was scheduled to pass diagonally over the at-grade intersection of two eight-lane arterial roadways. Standard construction techniques would have required the partial or full closure of travel lanes over extended periods of time, seriously aggravating an already unacceptable level of traffic congestion. Through the use of construction techniques more commonly applied to major structures, the arterial roadways remained open throughout the construction period and the construction cost was below the original Engineer's Estimate.

Traffic Management during Interstate Reconstruction (Syracuse, New York, USA) - During reconstruction and widening of a 10-mile section of interstate highway, a comprehensive and aggressive programme was undertaken to manage traffic congestion. Traffic signal improvements were made at 30 locations on parallel routes. Five new park-and-ride lots were opened. For each lot, six express buses were provided per peak period. Over 70,000 brochures were printed and distributed detailing alternative travel routes and modes (for a metropolitan area with a population of less than 500,000). The result was a 17 per cent reduction in peak period traffic in the corridor under reconstruction.

Highway Reconstruction Project in Pittsburgh (Pennsylvania, USA) - For a complete description of the project, refer to Section III.4.8.

Selected references

1. BIIOURI, Neila (1991). *A Comparison between Two Ramp Metering Strategies in Application on the Boulevard Peripherique in Paris.* RTS, March 1991. Paris.

2. COHEN, Simon. *Ingénierie du trafic routier.* Paris, Presses de l'Ecole Nationale des Ponts-et-Chaussées. Paris.

3. DREIF (1992). *Les transports de voyageurs en Ile-de-France 1991.* Direction Régionale de l'Equipement de l'Ile-de-France. Paris.

4. HADJ SALEM, H., et al. (1988). *Un outil de régulation d'accès isolé sur autoroute : étude comparative sur site réel.* Rapport INRETS.

5. INSTITUTE OF TRANSPORTATION ENGINEERS (1980). *Energy Impacts of Urban Transportation Improvements.*

6. IVHS America (1992). *Strategic Plan for Intelligent Vehicle-Highway Systems in the United States.* Washington D.C.

7. IVHS America (1992). *Guidelines for ATMS.*

8. OECD (1992). *Intelligent Vehicle-Highway Systems, Review of Field Trials.* Paris.

9. OECD (1989). *Traffic Management and Safety at Highway Work Zones.* Paris.

10. TRANSPORTATION RESEARCH BOARD (1986). *Transportation Management for Major Highway Reconstruction.* Special Report 212. Washington DC.

11. U.S. DEPARTMENT OF TRANSPORTATION (1980). *Traffic Control System Improvements: Impacts and Costs.* Washington D.C.

12. U.S. DEPARTMENT OF TRANSPORTATION (1983). *Freeway Management Handbook.* Washington D.C.

13. U.S. DEPARTMENT OF TRANSPORTATION (1990). *Operation Green Light, Annual Report.* Washington D.C.

14. U.S. DEPARTMENT OF TRANSPORTATION (1992). *Intelligent Vehicle-Highway System (IVHS) Projects in the United States.* Washington D.C.

15. U.S. DEPARTMENT OF TRANSPORTATION (1991). *Guidelines on the Use of Changeable Message Signs.* Washington D.C.

16. U.S. DEPARTMENT OF TRANSPORTATION (1989). *Ramp Metering Status in North America.* Washington D.C.

II.8. PREFERENTIAL TREATMENT

Description - Preferential treatment measures constitute any action taken to improve the travel speed, safety, or reliability of a particular mode of travel [note: measures which provide a financial incentive to use a particular mode of travel are found in Section II.5. **Economic Measures**]. This preferential treatment provides incentives that many people find attractive enough to change from driving alone to riding the bus, carpooling, bicycling, or walking.

The types of preferential treatment measures described in this section include:

♦ bus lanes,
♦ carpool lanes, also known as HOV lanes (for high-occupancy vehicle, e.g., a vehicle with 2 or more passengers, including buses),
♦ bicycle/pedestrian facilities,
♦ traffic signal pre-emption.

Most successful preferential treatment projects incorporate supporting services and policies as well as physical improvements, such as park-and-ride lots, bicycle shelters, and transit centres. Supporting services and policies may include new or expanded bus services, carpool and vanpool programmes, appropriate parking policies, employer supported transit-pass subsidies, marketing, and public relations programmes.

Bicycle-only lane with its own special traffic signal (Oslo, Norway)

Objectives and Major Impacts - The objective of implementing preferential treatment measures is to achieve a more equitable share of road space between different transport modes in order to meet public goals. For exclusive bus-lanes, the objective is to maximize the person-carrying capacity of a roadway or corridor. The benefits of travel time savings and more reliable transit travel times provide incentives for travellers to choose buses, rather than drive alone. In addition, exclusive bus lanes allow buses to operate at faster and more predictable speeds, thereby increasing bus productivity and reducing operating costs. For example, the bus lanes in Lyon, France, save an estimated five buses in the vehicle fleet during the peak period. For exclusive pedestrian or bicycle facilities, the primary function is usually to separate those travel modes from incompatible motorised vehicles in order to improve safety and often convenience.

Experience with preferential treatment facilities has shown that they provide benefits many travellers find attractive enough to change from driving alone. For example, an unusually high 90 per cent of the bus riders on an operating HOV facility in the U.S. have an automobile available, but choose to use the bus. Travel time saving is a key benefit for travellers. A travel time savings of between 10 and 15 minutes is a typical benefit of successful HOV preferential facilities. In addition, surveys of HOV lane users have indicated that the travel time reliability provided by the facilities is as important as the travel time savings.

Express center lanes for the use of buses and carpools (high occupancy vehicle lanes)
(Washington DC, United States)

For short distances, bicycling and walking are effective alternatives to driving alone. However the percentage of car-trips shorter than 5 km is considerable. In The Netherlands, where already 28 per cent of all trips are made by bicycle, still about 40 per cent of the car-trips are shorter than 5 km. In cities with traffic congestion, bicycle-trips at these short distances are often faster door-to-door than are car-trips.

HOV facilities can also enhance the productivity of buses and reduce operating costs. In otherwise congested freeway corridors, separate bus lanes can allow peak-hour bus operating speeds to double on average to free flow speeds. This results in significant reductions in schedule times for buses using the lanes and operational cost savings.

Application of Measure - A variety of preferential treatments are currently in operation throughout the world. Within the U.S., for example, there are approximately 42 HOV operations on freeways or in separate rights-of-way, accounting for some 500 km. In Paris alone there are 193 km of bus lanes. In addition, bypass lanes at freeway entrance ramp meters, arterial street bus lanes, and downtown transit malls are used in many areas. Further, additional projects are in the planning, design, and implementation process.

Good walk/bike facilities, along with higher density, mixed use development and good public transport service, can be considered part of a tripod. The three legs together support all alternatives to travel by the single occupant vehicle; the removal of any one of the three legs will weaken response to most alternatives to driving alone.

Several larger Swiss cities (Bern, Zurich) have applied special, permanently marked bus/taxi lanes. Also in Bern, Zurich, and Basle, there are special permanent marked bicycle lanes (with bike detectors at some signalised intersections)

Signal pre-emption for buses and trams is widespread throughout Europe. In 1989, 1600 signalized intersections and 3800 public transport vehicles were so equipped in cities in France alone. In almost all major German and Swiss cities, buses and trams are equipped with signal pre-emption equipment. In the Ruhr area, average transit vehicle travel speeds have increased 20 per cent and transit travel time has decreased 15 per cent with the initiation of signal pre-emption. Overall, it is estimated that 90 per cent of the Swiss trams and buses do not stop at these controlled intersections and operate for the most part at a route speed of 40 km/h. However field observations have indicated that the average speed of the remaining traffic can sometimes be worsened drastically, especially in the vicinity of major intersections with significant levels of public transport on all intersection approaches.

Institutional Responsibility for Implementation - A variety of institutional arrangements have been used to develop and operate preferential treatment facilities. In almost all cases, the responsible highway agency has been the lead agency in planning, designing, and constructing motorway facilities. Transit agencies have also been involved in many of the different bus lane projects.

Responsibility for the operation of the preferential treatment facility also is often spread among agencies. Generally, the highway operating agency is responsible for maintenance, although this may be a shared responsibility with the local public transport agency. Public transit operators usually are responsible for the provision of bus service and rideshare programmes, maintaining park-and-ride facilities, and other support programmes. Enforcement activities fall under the jurisdiction of the traditional police functions although, in some cases, responsibility rests with transit police. Thus, one of the unique characteristics of preferential treatment facilities is that a good deal of interagency

Specific traffic lights offer quick passage for buses at junctions

Special bus lane in a suburb of Rotterdam

cooperation and coordination is needed to ensure a successful system. This can be both an advantage and a limitation of preferential treatment projects.

Effects on Travel Patterns - Experience with the use of preferential treatment facilities indicates that they can influence travellers to shift to alternative modes. For new bus lanes and HOV facilities, surveys indicate that as many as 50 per cent of the bus riders and 60 per cent of the carpoolers previously drove alone.

Cost-Effectiveness - It is difficult to determine the exact cost effectiveness associated with the different types of preferential treatment projects. This is due in part to the problem of isolating the costs associated with preferential treatment facilities when they are part of a larger motorway or roadway project. However, preferential treatment facilities appear to provide a cost effective approach when properly designed and operated. In terms of capital costs, bus-only lanes and HOV lanes provide two advantages. First, when compared to other fixed-guideway transit projects, these projects generally fall toward the lower end of the cost spectrum (except when "separate" guideways are needed); second, a variety of funding sources are often available to be used to finance the preferential treatment projects.

Special Problems or Issues - Several issues are important to note in connection with the development and operation of preferential treatment projects. The most common concerns deal with HOV facilities and their vehicle occupancy requirement (or carpool definition) and enforcement. The optimum vehicle occupancy level is one that will produce a relatively high volume of use, avoiding what has been termed the "empty lane syndrome", but that also maintain conditions on the HOV lane that provide travel time savings and travel time reliability. One of the advantages of HOV lanes is that the vehicle occupancy levels can be changed to respond to changes in demand. For example, some facilities have reduced the occupancy requirements to encourage greater use of the facilities. In turn, the peak-hour occupancy requirement need to be increased, when the HOV flows begin to compromise the travel time savings and travel time reliability.

Enforcement of the vehicle occupancy requirement is important to maintain the integrity of the HOV facility. As pointed out, use by vehicles without the proper number of occupants reduces travel time benefits and leads to congestion in the HOV lane. Also, if drivers in the general-purpose lanes feel the HOV lane is being used by vehicles not having the proper number of occupants, they will probably give less support for the facilities. Thus, enforcement is a major concern. Experience indicates that addressing enforcement needs and concerns in the planning and design stages is important. Further, maintaining an ongoing enforcement programme is important. A variety of approaches are used to enforce HOV facilities, including: stationing enforcement personnel along the lanes, using special programmes, and experimentation with the use of advanced technologies.

Implementation of any preferential treatment on new right-of-way may also introduce negative reactions by adjacent property owners who have limited access to the facility but could experience what they perceive to be adverse impacts (e.g. noise from buses). As with any transport system improvement requiring the acquisition of new right-of-way, careful attention must be paid to the concerns of all affected parties.

Finally, implementation of any preferential treatment (e.g. bus, bicycle) should be carefully planned in order to not directly and adversely affect the other travel modes (e.g., auto, pedestrian) for fear of any negative backlash reaction.

Bus Lanes

Use of Hard Shoulders by Buses (The Netherlands) - In 1991, the Dutch Ministry of Transport and Public Works published a guideline concerning the use of hard shoulders by buses. This innovative measure is considered as temporary and only if other measures are not feasible. The number of bus passengers did not increase, however bus travel times did improve.

Tidal Flow Lanes for Buses (The Netherlands) - A tidal flow lane is usually situated between lanes of opposite directions and is available for the direction with the highest flow-rate. Typical effects include an increase in peak hour traffic volumes with no change in the daily volume. Queue lengths do not increase and average vehicle delay dropped significantly.

Nicollet Mall (Minneapolis, USA) - The Nicollet Mall in downtown Minneapolis was one of the first and most successful transit and pedestrian malls in the country. Opened in 1967, the mall has recently undergone a major renovation. The mall provides travel time savings for buses in the downtown area and provides a focal point for transit services.

Broadway/Lincoln Corridor (Denver, USA) - Special bus lanes are in operation on Broadway and Lincoln approaching downtown Denver. The lanes, which are approximately 4 km in length, are operated during the peak-periods. Approximately 117 buses, carrying some 4,800 passengers, use the lanes during the peak-period in the peak direction. These are mostly express and regional buses. The lanes provide travel times savings ranging from two to five minutes for buses in the corridor.

Bus Lanes in France - Application of bus lanes is widespread in France. There are 193 km of bus lanes within the Paris region, and 262 km in the provinces. As mentioned, in terms of effectiveness, the 550 m bus lane in Lyon saves an estimated five buses during each peak hour.

Bus Lanes in Germany - Special lanes have been constructed for line-haul public transport in many cities. At the end of the bus-lanes and at most of the signalised intersections along the bus lanes there are signal pre-emptions for the buses. Therefore, the considerable improvements in speed and reliability result from a combination of these two measures.

Carpool Lanes

Houston (Texas, USA) - Currently, 47 miles of a planned 96 mile system of exclusive HOV lanes are in operation in Houston. The HOV lanes are primarily one lane, reversible, barrier-separated facilities located in the median of the freeway. HOV lanes are in operation in four radial freeways, under construction in a fifth, and in the design stage on a sixth. The Houston HOV network also includes an extensive system of park-and-ride lots, express bus services, and direct access ramps. The HOV lanes have been effective in attracting new bus riders and carpoolers.

Seattle, Washington (USA) - Currently, approximately 49 miles of HOV lanes are open and another 49 miles are under construction. The network includes a downtown bus tunnel, exclusive barrier-separated HOV lanes, concurrent flow HOV lanes using both the inside lanes and the outside shoulder, and arterial street HOV lanes. Park-and-ride lots and improved bus services have also been implemented to support lanes. The HOV facilities have resulted in increased transit ridership and increases in carpooling and vanpooling. Currently 12 HOV bypass lanes are in operation at metered freeway entrance ramps in the Seattle area and additional bypass lanes have been proposed. The identification of the HOV bypass ramps has been a joint effort of the Washington State Department of Transportation (WSDOT), Seattle METRO and other transit agencies in the region, and local communities.

HOV Lanes on Arterial Roads (Sydney, Australia) - HOV lanes on arterial roads have been in operation since 1974. Travel time surveys have found savings achieved for vehicles travelling in HOV lanes ranging between 22 and 40 per cent. Average vehicle occupancy on HOV lanes is between 1.76 and 2.66, compared to 1.18 or less on normal lanes. The primary difficulty associated with arterial HOV lanes is enforceability. For example, illegal use of Sydney HOV lanes on arterial roadways ranges between 23 and 53 per cent.

Bicycle and Pedestrian Facilities

Bicycle Network in Delft (The Netherlands) - An integrated network of bicycle facilities is provided, including a city, district, and neighbourhood network. Facilities include cycle paths, underpasses and overpasses. The effect of the programme has been an increase in bicycle usage of as much as 8 per cent. There were shifts in the observed modal split of trips in the city (primarily from transit to the bicycle mode). Bicycle usage increased from 40 per cent to 43 per cent. The shares of both walking and car usage remained at 26 per cent. Public transport usage decreased from 6 to 4 per cent.

Bicycle Parking and Pathways (Denmark) - Four bicycle routes were constructed by the Danish government in four towns in 1984/85. The project was initiated to encourage people to use bicycles for their daily commute and to improve bicyclist safety. The routes offered continuous paths from residential areas to the city centre. There was no measurable effect on the corridor mode splits. To both pedestrians and motorists, their perception of conditions worsened. In terms of parking for bicyclists, covered spaces are provided at several local and regional train stations. Some spaces are located behind a locked gate requiring the use of a key, providing security for the bicyclist. At some locations privately operated bicycle centres have been established. These centres are manned throughout the day and offer the possibility of maintenance or repair of bicycles. The centres also often offer bicycles for rent.

Bicycle Facilities in Switzerland - Special and permanent marked bicycle lanes (or combinations with bus/taxi lanes) are in place in many Swiss cities (e.g. Bern, Zurich, Basle). These lanes are often equipped with special bicycle detectors which allow for preferential treatment at traffic signal controlled intersections.

Pedestrian Flow Improvements at Intersections (London, UK) - As part of the Red Route Pilot Project, the Traffic Director of London identified locations where there was a high number of accidents involving pedestrians. At one location, along a major shopping area on the main pilot demonstration route, large numbers of pedestrians were forced to cross six lanes of traffic. Traffic signals were timed to the pedestrian movements, thus creating long delays for motorists. On-street parking was also a problem at this location. The Traffic Director, working with local business owners and transport officials created new intersection designs that "neck-down" the wide intersection and only allow two lanes of traffic to flow in each direction. The improved design allowed for a safer and shorter distance for pedestrians to cross the road, better protection for parked vehicles using the former traffic lane, slower vehicle speeds, and more efficient signal timing created by shorter pedestrian crossing distances. As a result of this action, it is estimated that the 70 pedestrian accidents that occurred over the period 1989 to 1992 will be reduced to 30 pedestrian accidents during the period 1993 to 1996. Because of the success of this type of project, similar pedestrian and bicycle friendly improvements were installed at other key intersections along the Red Route Pilot Project.

Traffic Signal Pre-Emption

Traffic Signal Pre-Emption by Buses (France) - Roughly 3,800 vehicles and 1,600 traffic signals have been equipped with transmitters and receivers, respectively. For one line in Paris with headways of just over two minutes, the estimated time saved per bus is 10 seconds per equipped intersection. A survey showed a gain of 3 per cent in the average speed of buses.

Traffic Signal Pre-Emption by Buses and Trams (Switzerland) - In almost all Swiss towns and cities (e.g. Bern, Zurich, Basle), public transport vehicles (buses and trams) are equipped with special devices (e.g., FM- or infrared controlled transmitters or special detectors) and are able to change traffic signals and thus obtain priority.

> Traffic Signal Pre-Emption by Buses and Trams (Germany) - There is widespread use of public transport signal pre-emption throughout Germany. As an example, the results of before-after studies for different signal pre-emptions in the Rhein-Ruhr area found an average increase of speed of about 20 per cent which reduced travel time by 15 per cent.
>
> Traffic Signal Pre-Emption by Trams (Melbourne, Australia) - Active tram priorities have been in operation at traffic signal controlled intersections in Melbourne since 1984. At present all 450 signalized intersections on tram routes operate with special tram priority movements. Priority is given primarily to trams travelling in the peak direction. The ability to give priority to trams has been provided by the computerized signal control system, SCATS.

Selected references

1. U.S. DEPARTMENT OF TRANSPORTATION (1992). *A Description of High-Occupancy Vehicle Facilities in the United States*. Washington D.C.

2. U.S. DEPARTMENT OF TRANSPORTATION (1991). *Suggested Procedures for Evaluating the Effectiveness of Freeway HOV Facilities*. Washington D.C.

3. U.S. DEPARTMENT OF TRANSPORTATION (1991). *An Evaluation of the Houston High-Occupancy Vehicle Lane System*. Washington D.C.

4. U.S. DEPARTMENT OF TRANSPORTATION (1990). *HOV Project Case Studies - History and Institutional Arrangements*. Washington D.C.

5. U.S. DEPARTMENT OF TRANSPORTATION (1989). *Planning, Design and Maintenance of Pedestrian Facilities*. Washington D.C.

II.9. PUBLIC TRANSPORT OPERATIONS

Description - A variety of public transport (transit)[1] operations and services are currently provided in most metropolitan areas and communities. The expansion of public transport service is usually considered an important component of an urban congestion management programme because by shifting persons' travel from automobiles into public transport vehicles, regional traffic congestion will be reduced.

Public transport operations measures include the following measures:

♦ express bus lines

---

[1] Terminology - English: public transport; American: public or mass transit.

- ◆ park-and-ride facilities
- ◆ service improvements
- ◆ public transport image
- ◆ higher public transport vehicle capacities

Express buses offer fast, frequent, and convenient service with a limited number of pick-up and drop-off points. Interurban coaches are often used and premium fares may be charged. Park-ride facilities are sites where motorists can park an auto and await the arrival of public transport. Service improvements encompass traditional coverage expansion and service frequency changes. They also include institution of a wide variety of non-traditional public transport services such as circulator services which link residential and employment or other areas (e.g. recreational) with rapid transit systems and mid-day services linking employment areas with restaurants or services in nearby commercial areas. Another service improvement would be deeply discounted fares. For some years, several major cities in Greece offered free public transport before 0800 in order to encourage their usage prior to the morning peak period.

Transit-only Zone (Oslo, Norway)

Cooperation between public transport operators on tariffs and schedules has been a major point of concern in the Netherlands. Schedules were coordinated between bus services and train services and tickets for any travel mode were made purchasable from any public transport operator. In addition, several Swiss towns and regions (e.g., Berne, Zurich, Basle) offer combined tickets which include all public transport providers, both within the region and throughout Switzerland.

71

Objectives and Major Impacts - The basic objective of implementing public transport service improvements is to provide transit mode alternatives that individuals will find attractive enough to cause their switch from the drive-alone mode. Thus, the focus of public transport service improvements is to provide travel time savings, cost savings, convenient services, and other amenities that individuals will find attractive. Funding constraints have dictated that another key objective of public transport service improvements is to provide greater opportunities for private sector involvement in the planning, operating, and financing of transit services.

Park-and-Ride facility with provisions to accomodate bicycles (Münster, Germany)

Application of Measure - Public transport service improvements for the purpose of reducing traffic congestion make sense only if the service will generate sufficient patronage.

In addition to traditional service expansions, a number of innovative public transport operations have been implemented throughout the world. Some of these have been undertaken as limited demonstration projects, while others have been introduced more widely. Some of the projects have been implemented relatively recently. As a result, many of the impacts and effects are just beginning to be examined and evaluated. Thus, in some cases, the effects of the different techniques are not well known. Examples of the different applications include contracting with private carriers to operate park-and-ride express bus service, reverse commute buspools to major employers, and shuttle bus connections between major employment centres and the regional heavy rail public transport system.

Express bus services operate traditionally on radial routes to downtown areas (see Stockholm, Minneapolis, Houston for examples). They are most effective when used for both long trips (over 25 km) and short trips (e.g. from peripheral park-and-ride lots).

Service improvements can include the traditional route extensions, headway reductions, and hours of operation expansions; all-day services circulating through residential areas (generally using vans or other small vehicles); peak-period circulator service between residential/employment areas and regional rail or bus transit stations/stops or park & ride lots; and mid-day circulator services linking employment and commercial uses.

Institutional Responsibility for Implementation - A variety of groups are usually involved in the provision of traditional or innovative public transport operations. The number and types of groups and agencies depend on the type of strategy or service application. The local public transport operator obviously leads any expansion of current transit services. For the innovative strategies described above, the public transport operator may be responsible for actually planning, implementing, and operating the service; it may contract with a private provider to operate the desired service; or it may provide technical assistance to private sector groups or employers who are responsible for implementing and operating the service. Private sector operators and businesses may be involved in a number of different ways. First, private operators may have the opportunity to bid on services to be provided in certain areas. Second, employers working alone or in groups may develop services for their employees. These services may be implemented in response to identified needs or to specific government regulations.

Effects on Travel Patterns - Experience has demonstrated that growth in ridership typically requires commensurate service improvements. For example, to increase ridership a certain per cent, an equal or somewhat greater per cent increase in bus-km of service is needed. The specific impact of public transport service expansion which is of interest here is the reduction in congestion which results from a decrease in total vehicular travel. All of the public transport operations improvements described here have direct positive impacts on overall regional mobility. For example, the targeted express bus service cited above has been found to remove nine times as many kilometres of single-occupant vehicle usage as are added in transit service.

The extensive experience which the public transport industry has with the application of different public transport operations improvements indicates that they can be an effective approach to encouraging travellers to change from driving alone. Operation of park-and-ride express bus service by private carriers has attracted new riders to the bus system and at a lower cost for the operator. Reverse commute buspool programmes have enabled the retention of employees for a relocated employment site and expanded the employment market for area residents.

Cost-Effectiveness - The cost effectiveness of public transport removing automobiles from the roadway network is highly dependent on the details of the particular setting. Basically, the cost effectiveness of any public transport scenario is proportional to the service's ability to attract riders and the tradition and culture of the city. In the simplest of terms, if enough riders can be gained to fill almost all of the seats in a transit vehicle, then this service would result in a respectable cost per automobile removed from the system. The limited experience with the different innovative public transport operation applications indicate that they can provide cost effective approaches.

Special Problems or Issues - A substantial expansion of public transport services can be an extremely complex institutional undertaking. As with other major public capital expenditures which are to be paid for by multiple jurisdictions, implementation of any measures requires long lead time, complicated multijuridictional planning and programming, and often thorny political negotiations. For example, the project may represent the first or one of the first applications of a specific technique or in a previously-unserved area and the ability to learn from other projects may not be possible. Further, planning techniques may not be fully developed to estimate the demand for many of these services.

Determining who will fund the service and who will operate it may be issues in some areas. If the service is focused on a specific group, such as a major employer, the public sector groups or the local public transport operator may try to obtain financial support from this business. This is especially true given the limited budgets and financial constraints facing many public transport operators today. The role of public and private sector providers may also be an issue. Labour provisions and union contracts may limit the ability of a public transport operator to contract with private operators for service.

Finally, many areas are difficult to serve with transit. The lack of "transit friendly designs" and the need for individuals to have access to automobiles for midday trips make serving many "newer" developments difficult. Further, the high levels of automobile ownership in suburban areas make it more difficult to attract choice riders to transit.

Examples

Express Bus Services

Premium Bus Service in Amsterdam (The Netherlands) - Luxury shuttle buses are used on routes with a high number of commuters. The service provides each passenger a guaranteed seat, newspaper, coffee, and separation between smokers and non-smokers. The seats are upgraded in comfort and the buses seat only 32 rather than 40 in order to provide more room per passenger. The service uses the existing bus lanes and traffic signal pre-emption. The reported previous mode of travel for the premium service users is as follows:

- ◆ 41 per cent by other public transport
- ◆ 39 per cent by private automobile
- ◆ 6 per cent by a combined car/transit trip
- ◆ 6 per cent in a vehicle provided by employer
- ◆ 5 per cent by bicycle or motorbike
- ◆ 3 per cent by carpool.

I-35W Express Bus Service (Minneapolis, USA) - Between 1971 and 1974, the Metropolitan Transit Commission (MTC) implemented twelve 35W express routes as part of the Bus-on-Metered-Freeway Demonstration Project. These routes provide peak-period express service from suburban communities into downtown Minneapolis. The service is oriented toward park-and-ride-lots with, some limited neighbourhood stops. Approximated 11,000 daily riders use this service. This service has kept a significant number of automobile off of I-35W.

The Metropolitan Transit Authority of Harris County Park-and-Ride Service (Houston, Texas, USA) - In conjunction with the development of the Houston HOV lanes, Houston METRO implemented an extensive system of park-and-ride lots and service. Buses operate from the park-and-ride lots, providing express service to the downtown area and other major activity centres. The service and the HOV lanes have attracted large numbers of new riders. Further, METRO has been able to realise cost savings by contracting with private operators for some of the service.

Park-and-Ride Facilities

Peripheral Park-and-Ride Facilities in Switzerland - In many Swiss towns (e.g., Berne, Basle, Zurich), peripheral park-and-ride facilities with free parking privileges and express bus services and/or good railway connections have been constructed.

Service Improvements

SAE Bus Transport Operation Systems (France) - In France, bus transport operation systems (SAE) have been developed. A communication system allows an automatic exchange of data in real time between vehicles and a central operation place. It results in an improvement of prefixed schedules and real time operation. More than 40 public transport networks are equipped with SAE (for a total of 3,150 vehicles and 350 lines). In Nancy, an improvement of 4-5 per cent was observed on the journey times and 21-72 per cent of the average delay at terminus arrivals. The global measured productivity gain was 11.8 per cent.

Roseville Area Circulator Service (St. Paul, Minnesota, USA) -In 1989, the Regional Transit Board (RTB) implemented the Roseville Area Circulator service. The service is comprised of small buses operating on five routes in the northern suburbs of St. Paul, Minnesota. The service is provided by a private operator, under contract to the RTB. The routes meet at a regional suburban shopping centre, allowing passengers to transfer to regular route service to downtown Minneapolis and downtown St. Paul, and between the Circulator routes. The vehicles are equipped with wheelchair lifts and can accommodate bicycles. The service has been well received and the routes have been expanded.

Tyson's Corner/Fair Lakes Shuttle (Virginia, USA) - Tyson's Corner and Fair Lakes are two suburban office, retail, and residential complexes in Virginia outside of Washington, D.C. Both organized shuttle bus services to the Metro rail stations in their area. The shuttle service provides a link to Metro services for employees, shoppers, and residents. In 1991, the Tyson's Corner shuttle was carrying an average of 300 passengers on an average weekday. The service had a operating cost of $1.29 per passenger and a revenue/cost ratio of 39 per cent. The Fair Lakes shuttle was averaging 240 daily riders.

Summer Bus in Zeeland (The Netherlands) - Bus service was initiated during the summer months in order to decrease the use of automobiles by recreational visitors to the region. Bus routes pass several camping sites, recreational parks, the beach, and several other tourist attractions. The programme resulted in a relatively modest reduction in total vehicle usage, but improved the mobility of area visitors.

Red Routes in London (U.K.) - In 1991, a system of 500 km of priority (Red) routes were designated. The programme includes the designation of loading zones, provision of sufficient short-term parking spaces, and allocation of space for peak period bus lanes. The dramatic results of an extensive before-and-after pilot test are as follows: bus travel times were reduced over 20 per cent and bus on-time reliability improved 33 per cent; bus patronage increased 3 per cent; motorist accidents declined 17 per cent; and overall travel time for all vehicles improved over 20 per cent. The only exemptions to red line restrictions are for buses, for licensed taxis picking up or dropping off passengers, for handicapped person pick-up or drop-off, and for emergency purposes.

Poolective Transport (Sweden) - In two villages outside Goteborg located along railway service, 8-passenger vans were operated to both serve as a shuttle to the railway station and to accommodate other trips such as for daycare and other children-related functions. The drivers were recruited on a voluntary basis among the residents and operated the bus on a non-paid basis. The mileage within the area was decreased and the need for a second car diminished. Similar operation is known in the Netherlands as *Buurtbus*.

> ### Public Transport Image
>
> Public transport operators in France have undertaken significant efforts to upgrade the image of public transport. Measures taken have addressed vehicle decoration (internal design and exterior painting), cleanliness of buses and stations, amenities at stations and bus stops, in-vehicle amenities (audio systems, music on buses), enhanced security (alarm systems), accessibility of vehicles for disabled persons, and improved and targeted advertising.
>
> ### Higher Public Transport Vehicle Capacities
>
> Public transport operators in France have begun experimentation with the Megabus, an articulated bus with three bodies. Preliminary tests of vehicle mobility, loading/unloading times, and effects on other traffic have all been positive.

Selected references

1. U.S. DEPARTMENT OF TRANSPORTATION (1993). *Introduction to Public Finance and Public Transit.* Washington D.C.

2. U.S. DEPARTMENT OF TRANSPORTATION (1992). *Advanced Public Transportation Systems: The State of the Art.* Washington D.C.

II.10. FREIGHT MOVEMENTS

Description - Freight movements and services throughout urban and rural areas is a major activity in terms of both scale and importance and represents a key component of the problems requiring "congestion management" solutions. Basically, the speed, cost, and reliability with which goods and resources move within a region often define the potential economic health of that region. In quantitative terms, the U.S. Department of Transportation estimates at the national level that 25 tonnes of freight are moved annually per capita. At the local level, freight tonnage destined for direct consumption by New York metropolitan area residents was estimated 30 years ago to be more than 5 tonnes annually per capita. And in the past few decades, there has been tremendous growth in the volume of freight moved, in the number of small shipments, in the use of private carriers, and in the dispersion of development -- all of which exacerbate the need for efficient freight movement.

The International Road Federation (IRF/FRI) has estimated that the proportion of freight (in tonne-kilometres) moved by truck is 80 per cent in the United Kingdom; between 50 and 60 or more per cent in France, Germany, and Japan; and roughly 30 per cent in the United States. Therefore, freight movement by truck both within an urban area (i.e. goods movement, deliveries) and inter-city is an issue. The OECD Report (1992) on "Cargo Routes: Truck roads and Networks" presents a comprehensive review of traffic management measures (and infrastructures needs) and discusses evaluative criteria for requirements of separating trucks and automobiles taking full account of capacity and congestion issues as well as safety and environmental goals.

The problems preventing this efficient movement and delivery of freight by truck are many. Buildings, especially those in dense urban areas, frequently do not have off-street loading facilities and must rely on available on-street curb space. Even if off-street facilities are provided, they may not be navigable by larger trucks and they may have insufficient bay space. Roads leading to delivery destinations are sometimes too narrow (e.g. for passing opposing traffic, for turning) or of inadequate design to accommodate heavy truck axle-loadings. Other elements of the infrastructure can also be impediments, such as low overpass clearance, utility poles and traffic signals located too close to the curb, and street furniture.

Objectives and Major Impacts - The overriding objective of freight movement measures is to facilitate transport operations. Each stage of the delivery process can contribute to the efficient (or inefficient) movement of goods:

♦ the movement of delivery vehicles in the traffic stream;
♦ the location, quantity (i.e. availability), and size of loading zones of the shipper and receiver of the delivery, which enable efficient parking;
♦ the proximity of parking to delivery spot which reduces dwell time; and
♦ the readiness of the receiver of delivery to load/unload freight, check shipment, etc.

Any improvements made to general traffic flow (such as those described in Section II.7., **Road Traffic Operations**) will likewise reduce travel time for delivery vehicles. Truck-oriented improvements such as removal of operational and physical constraints will reduce delays to truck traffic and therefore to general traffic flow. Restricting trucks from certain roadways will reduce conflicts between trucks and autos and pedestrians, will reduce road maintenance costs, and could improve overall traffic flow. Likewise, the provision of truck-only roads, ramps, or interchanges could reduce truck travel time, reduce auto/truck conflicts, reduce truck volume on other streets, and improve the homogeneity of traffic on facilities.

Improvements to loading areas (both on-street and off-street) will reduce search time for carriers, improve driver productivity, reduce double parking, and increase turnover of available space for loading and unloading.

In the long-term, measures such as off-street truck terminals and automated goods transport could likewise have significant positive impacts on goods movement efficiencies and costs.

Application of Measure - The private sector can take a number of steps to facilitate loading and unloading functions at building sites:

♦ Improvements to off-street loading areas to better accommodate size and configuration of trucks;

♦ Changes within buildings to facilitate movement of freight (e.g., upgrade levators in older buildings);

♦ Improve shipping logistics through just-in-time delivery, consolidation of shipments, and simpler documentation (to reduce paper work and save time during delivery);

♦ Create freight consolidation rooms on the ground level of large buildings to reduce turn around time for delivery vehicles.

Combined transport (Germany)

Public sector actions to improve general traffic flow could likewise benefit goods movement. These include physical street system improvements such as improving the roadway alignment, one-way streets, traffic signal system improvements, intersection improvements (channelization; lane controls), peak period curb parking restrictions, and reversible lanes. Likewise, operational and physical constraints which particularly affect truck movement (such as traffic signals, turning radii, vertical clearance and height restrictions, median width, roadside obstacles) can be addressed by the public sector.

The proper allocation and management of curb space is critical to use of on-street loading and unloading. Policies should be enacted by both the public and private sectors to facilitate use of the curbside so that it can be efficiently shared among all legitimate users with due consideration of their competing interests. This allocation should be premised on vehicle priorities based on community needs with the primary criteria to include effectiveness, urgency, and available alternatives to use of the curbside. In addition, the curbside allocation policies should include objective warrants for prohibiting stopping, standing, or parking based on the needs of traffic flow and safety; loading zone and entrance requirements and practices; warrants for frequency and placement of bus stops and taxi stands; warrants for reserving curb space for tourbuses, vanpools, vendors, limousines, and other special users; and criteria for time limits on parking and standing.

Truck restrictions can address various elements of the goods movement universe: delivery times, through traffic prohibitions, compulsory routes, etc. For example, in Switzerland, trucks are not allowed to circulate during the night and on Saturdays, Sundays, and holidays (unless granted permission to transport perishable goods). France has similar restrictions on truck travel, for example during the summer months on weekends, holidays, and the days before holidays. In Paris, the times during which trucks can stop on-street is affected by their size. As the truck size increases, the permitted time periods shorten (e.g. the largest trucks can load/unload only at night between 9.30 p.m. and 7.30 a.m.).

Often the competition between on-street general purpose parking and the need for curbside loading and unloading is untenable. In those cases, the options of either providing additional off-street parking for the general public or of providing off-street loading facilities must be pursued.

In the longer-term, consideration should be given to implementation of truck-oriented roads (which could be open to all traffic but provided to primarily to serve truck travel needs), truck-only roads, and truck-only ramps or interchanges. (See 1992 OECD Report).

Institutional Responsibility for Implementation - The shippers and receivers of freight; the carriers; the designers and owners of building facilities; the providers of transportation infrastructure; and the enforcement agencies are all responsible for the implementation of goods movement measures.

The shippers and receivers (including property managers and owners) are concerned with the cost, quality, and convenience of freight-service; with the cost of providing shipping and receiving facilities, and with the costs associated with damaged or delayed shipments. There are numerous on-site measures solely within their purview.

The carriers (e.g. freight companies, delivery services) are concerned with the cost and delay associated with areawide and localized traffic congestion; the distances between shippers and receivers; inadequate design of shipper and receiver facilities and limited hours of operation; geometric design and traffic control along roadways; routing restrictions; illegally parked vehicles in loading zones; and security for shipments. As a primary beneficiary of goods movement measures, the carrier can take an active role in their implementation.

The provider of the transport infrastructure (generally the highway authority) is concerned with damage and deterioration of roads and bridges not designed to accommodate heavy trucks and with the need to provide for the safe and efficient movement of both people and goods. The provider has direct responsibility for all measures which require implementation on public road rights-of-way, including along the curb.

The benefits accrued to the enforcers of goods movement measures (e.g. curbside restrictions) are primarily indirect -- with adequate enforcement, other problems such as motorist safety and pedestrian safety are minimized. Compared to many of the other congestion management measures described throughout this chapter, goods movement measures are not as readily self-enforceable. Their effective implementation requires a commitment to substantial levels of enforcement.

Cost-Effectiveness - The U.S. trucking industry has estimated that the cost of delivering a tonne of freight in a central urban area is, on average, three times that of delivering the same tonne to a suburban location. In turn, suburban deliveries cost about twice as much per tonne as do those into rural areas.

The cost-effectiveness of requiring night deliveries is still debatable. The potential for added staff requirements and higher security risks has made night delivery an expensive proposition for the persons paying for the delivery of freight (which will in-turn be passed through to the general public). However, the general public may benefit on average with the associated reductions in daytime traffic congestion.

Special Problems or Issues - All truck loading zones need to be strictly policed in order to assure that none but legitimate freight carrying vehicles occupy them.

Loading zones are often used by such service vehicles as telephone and equipment repairers/installers, painters, electricians and other service personnel. They occupy the loading zones for extended periods of time while their actual freight usually consists of light-weight materials off-loadable in less than five minutes (which should be their maximum allowed dwell time). Following such off-loading, service vehicles should be placed at other on- or off-street regular parking spaces.

Loading zones should be clearly marked both on the adjacent sidewalk (by means of signs) and on the pavement. Signing should reflect applicable hours and maximum allowable dwell times (which should seldom be permitted to exceed 30 minutes). Placement of signs on city streets should recognize the need for balance between signing requirements to manage curb space and the associated aesthetic problems. From the perspectives of the shipper, receiver, and carrier, loading zones should be kept clear of potential curbside impediments as planters, benches, other signing, etc.

Freight movement often involves drivers in unfamiliar areas. Uniformity of laws and parking regulations between jurisdictions (e.g. the use of international symbols for parking/regulatory signing on city streets) will facilitate loading/unloading.

Time-of-day delivery restrictions often are not acceptable to the shipper or receiver. Especially in downtown business districts, merchants and restaurants, as examples, usually have little or no off-floor storage space. For fiscal, security and insurance reasons they cannot remain open in off-hours for deliveries. Therefore they must be served literally on demand (i.e. throughout the business day).

A differentiation should be accommodated between the growing number of small parcel and express delivery vehicles. These seldom require the same curbspace length as do large delivery vehicles and have little need to remain more than 10-15 minutes. Designation of a parcel vehicle zone during the peak delivery time periods (e.g. morning and evening peak periods) could be considered.

All new and substantially rehabilitated downtown structures should be provided with adequate and properly designed off-street loading docks or similarly adaptable facilities. The effect is to remove loading/unloading truck activities from the street.

All new and substantially rehabilitated multi-tenant office structures within a downtown business district should include a central freight receiving/dispatching facility. Costs, including operating costs, should be treated the same as any other code-required facility and should be borne by tenants. The effect is to improve the efficiency of urban trucks by reducing dwell times substantially.

Urban Goods Movement

France established national regulations in 1992 which addressed truck restrictions. These regulations include the prohibition of dangerous goods and hydrocarbons throughout most of the weekend and on days prior to public holidays and the restriction of other trucks to only a few hours on these same days. At the local level, additional restrictions have been instituted. In Paris, for example, vehicles with a surface under 12 square meters can load/unload throughout the day along permitted roadways; for surfaces between 12 and 16 square meters, evening peak period loading is prohibited; for surfaces between 16 and 20 square meters, loading is forbidden throughout the afternoon; and for surfaces greater than 20 square meters, loading is permitted only at night.

Curb Space Management in Nashville (USA) - After a comprehensive evaluation of parking needs, a series of key steps were taken to improve the delivery and pick-up of freight in the urban environment. Certified courier/parcel vehicles were permitted to use freight loading zone but for no more than 10 minutes. Service vehicles were prohibited from parking in loading zones for longer than 5 minutes, which was determined to be sufficient time to load/unload heavy equipment. Size, location, and hours of operation of loading zones were refined. Designated space was allocated to bus stops, taxi stands, and tour buses. Finally, management and enforcement of the public parking spaces was enhanced.

Underground Truck Terminal (Perth, Australia) - Perth constructed an underground truck terminal to service retail outlets in the downtown area. The central shopping precinct is a pedestrian-only area after 10:00 and goods may only be delivered in this area before 10:00 or from trucks using the truck terminal.

New York City Garment District (USA) - In order to improve the efficiency of deliveries in the New York City Garment District, an auto-free zone was created in roughly a six-block area. Between 1000 and 1500 on weekdays, all auto traffic without a destination within the designated area is prohibited. The estimated auto trip reduction has been 30 per cent.

Inter-City Goods Movement

Combined Transport (Highway and Rail) (Switzerland) - The "Agreement between the Swiss Confederation and the European Economic Community on the transit of goods by road and rail" came into force in January 1993 and promotes goods traffic through the Alps. It facilitates the transport of units (containers, swap bodies, semi-trailers) mainly by rail and then by road in the terminal areas. Presently, the modal breakdown of total goods movement through the Swiss Alps is 22 per cent by road, 52 per cent by rail, and 26 per cent by combined transport.

Intercity Truck Transport in Germany - In Germany, there are different truck restrictions on different levels. On the national level, there are driving prohibitions for all trucks with more than 7.5 tonnes on Sundays on all roads and on weekends during the summer holiday period on the auto-bahns. On the local level there is the possibility for city authorities to ban through-traffic either for trucks with more than 7.5 tons or for dangerous goods.

Selected references

1. OECD (1994). *Advanced Road Transport Technologies.* Japanese Ministry of Construction. Tokyo.

2. OECD (1992). *Cargo Routes: Truck Roads and Networks.* Paris.

3. OECD (1992). *Advanced Logistics and Road Freight Transport.* Paris.

4. OECD (1991). *Future Road Transport System and Infrastructure in Urban Areas.* Japanese Ministry of Construction. Tokyo.

5. U.S. DEPARTMENT OF TRANSPORTATION (1979). *Urban Transportation Planning for Goods and Services.* Washington D.C.

CHAPTER III

DEVELOPING CONGESTION MANAGEMENT PROGRAMMES

The previous chapter illustrated the diversity of the measures which have been taken in different OECD countries in order to manage road traffic congestion. Many of the measures that have been implemented to relieve congestion have positive effects. But often, congestion has continued spreading, because different factors have increased traffic demand faster than implemented measures have relieved existing congestion.

In order to win the race against the growth of traffic, implementation of the measures which have been mentioned may be developed in two ways. The first is to attempt to apply a particular measure systematically in a large number of situations. The second approach for developing the implementation of congestion management measures is to combine several measures in a particular case in order to accumulate benefits. In fact, a single measure is rarely sufficient to remedy the situation, and a set of measures must be introduced in order to cope with a particular congestion situation. By combining different measures it becomes possible to produce coherent congestion management programmes.

A "Congestion management programme" can be defined as any combination of actions which aims to apply one or more specific traffic demand or transport supply management measures to a region or locality on a systematic basis, or which attempts to deal with all aspects of a congestion situation at a given location.

The purpose of this chapter is to describe the implementation of such programmes in OECD countries and to address the important factors that can lead to effective congestion management. The first stage is to stress the relative importance of congestion management in transport policies. Next, the difficulties involved in implementing congestion management measures, and the scarcity of existing programmes, will be explained. Then different factors will be listed so as to contribute to effective congestion management programmes. Finally will follow a description of some of the programmes of this type which exist in different OECD countries.

III.1. CONGESTION AS TRANSPORT POLICY AIM

One fundamental reason for the scarcity of congestion management programmes in most OECD countries is as follows: congestion is only one of many elements which make up transport policy. Transport policy usually includes many aspects: budgetary resources, access provision and regional

Congested arterial along Red Route "before" improvement (London, UK)

Arterial along Red Route "after" enforcement of parking restrictions and bus lane improvements (London, UK)

planning, the responsibilities of different territorial levels, transport safety, environmental protection, resident protection movements, energy savings. Policies are frequently the result of a compromise between these various demands and between the different parties involved in negotiations. It is for this reason that road traffic congestion as such is rarely the subject of specific programmes.

However, programmes which are not primarily concerned with congestion may affect congestion; for example, improvements in air quality -- by reducing the amount of pollutant emission as in California --, or reduction in energy expenditure often involve a reduction in congestion. Also, the taxation policy can be of primary importance for some of the measures for congestion management.

In the following sub-sections, the relations between major transport priorities in some OECD countries and management of congestion will be shown.

III.1.1. Road safety

In several OECD countries, road safety is a national priority.

For example, in France, the number of accidents and casualties is distinctly higher than in other industrialised countries. Many measures have been introduced in order to eliminate this phenomenon (speed limits, compulsory wearing of safety belts, monitoring of blood alcohol level, inspection of vehicles of more than five years of age, point-system for driving licence), and others are in the process of being implemented or developed. These take the form of either local measures (lowering of speeds in built- up areas), or more comprehensive programmes. For instance, the REAGIR programme consists of a methodology promoted by the Ministry of Transport: each fatal accident has to be analysed by a local specialized committee to propose measures to avoid new accidents on the same place. The national "Cities safer" programme encouraged local authorities to define comprehensive strategies, including various measures, to improve quality of life and road safety in urban areas.

Other measures have been taken in Germany, such as the introduction of a 100 km/h speed limit on the congested parts of the network, the introduction of a 30 km/h speed limit in residential zones of urban areas, and measures to improve traffic flow on motorways (such as diverting traffic onto parallel motorway routes in the event of congestion by using additional variable message signs in the place of existing conventional signs), speed restrictions in the parts of the network which are frequently congested, fog and congestion warning systems (coupled with speed restrictions), and traffic advice through broadcast.

Road safety programmes such as these reduce the adverse safety effects of congestion, such as the risk of rear collisions when unprecedented queues of traffic develop, but also, as can be seen in the case of Germany, reduce congestion itself.

III.1.2. Environment and reductions in vehicle travel

The environment is a major priority in transport policies in many countries. It often leads to modes other than road transport being promoted, both for personal transport (by favouring the use of public transport or bicycles to the use of passenger cars) and freight transport (by using water, rail, or combined transport in order to reduce heavy vehicle road traffic).

In Switzerland, political reasons, which are essentially related to environmental protection as well as growing opposition to expand the road network, make further construction in the country virtually impossible, with the exception of the trunk road system which needs to be completed. The future will consist more of improving weak points and managing the existing system as well as possible by using new information systems so as to achieve a uniform distribution of traffic over the existing network. In return, high priority is given to increasing the capacity of rail transport, which is considered as an environmentally friendly mode of transport, particularly in view of the increased traffic generated by the European Community. It is therefore also necessary to modernize public transport. For this reason, in the next few years major rail investment is planned, particularly on the main north-south axis (the New Transalpine Rail Line). The increase in rail capacity should indeed reduce road congestion. Thus, in Switzerland, an increase in freight and persons transported by rail and public transport means will result in a relative reduction of traffic on over-trafficked roads where there is a risk of congestion.

In The Netherlands too, environmental concerns are paramount. The Second Dutch Transport Structure Plan (1988-1990) planned to limit the growth of road traffic by a whole series of measures: planning relating to land use and the location of activities, telecommuting, car sharing for commuting trips, rail investment, improvement of public transport services, etc. Within this policy also a reduction of congestion is being pursued.

Congestion management programmes are also integrated into environment policies in the United States because of recognised relationship and impacts. For example, the Clean Air Act Amendments of 1990 will also require cities to implement congestion management measures (either individually or in groups) as part of their environmental programmes.

III.1.3. Land-use and zoning policy

In some countries, congestion management programmes are also integrated into land-use and zoning policies. In the United States, many cities are developing land-use policies to help encourage public transport, ridesharing, walking and bicycle activity, especially for the work trip. For example, in Montgomery County, Maryland, outside of Washington D.C., builders of new offices are required to implement and fund programmes that encourage their employees to use more efficient modes of travel to work. In addition, zoning laws have been changed to establish a maximum amount of parking that can be built rather than a minimum amount. This in effect places a limit to the number of parking spaces that can be provided at employment sites. Public transport, bicycle, and ridesharing measures are implemented in order to accommodate those that chose not to drive alone to work. Similar land-use and zoning measures are taken in the Netherlands, and parking limitations is the backbone of transport policy in London.

III.1.4. Current status and future perspective

While in most European countries, congestion management is often a part of a broader transport policy, in the United States congestion management programmes have been developed as a separate transport policy at the Federal, State and local levels. The States of California, Washington, and Texas for example have policies in place that have encouraged cities to develop coordinated approaches to congestion management. These programmes are intended to reduce the number of vehicles used for work trip travel, to reduce congestion during peak times, and to reduce air pollution.

In addition, new federal requirements for a Congestion Management System, under the 1991 Intermodal Surface Transportation Efficiency Act (ISTEA), will mandate that each State and urban area have a congestion management programme by 1995. The reason for this is that the U.S. Congress believed that congestion management is an important part of the national transportation policy and needs to be coordinated at the State and local level by the various agencies that have responsibility for improving traffic and transport operations. To ensure its importance, the U.S. Congress made all Federal transportation funding availability conditioned to a State having a Congestion Management System. So congestion management programmes are an important part of transport policy in the United States; however, it is not always obvious. Congestion management programmes may appear in other policies, such as environment, land-use, traffic management, safety, or freight movement.

Thus, in many OECD countries, road traffic congestion management programmes are often integrated within more general transport policies.

III.2. IMPLEMENTATION OF PROGRAMMES

Although it would seem essential to develop congestion management programmes, the task is not an easy one, as it is jeopardized by a number of obstacles, mainly institutional in nature.

Passing from isolated individual measures to genuine congestion management programmes may involve a number of intermediate steps. Policies may consist either in implementing a set of actions by a single party or in having several parties co-operate in order to apply one or several measures. Sometimes, co-ordination between several parties leads to the creation of a specific unit for the purpose of overcoming the institutional difficulties involved when different authorities work together.

Finally, comprehensive congestion management programmes, are the most perfected form of congestion policy. In view of the great many aspects which must be dealt with in most cases and of the large number of participants involved, which result in the need for institutional compromise, it is not surprising that systems of this type are relatively uncommon in OECD countries. Nevertheless, much can be learnt from the examples which exist about the approach which should be followed and the problems to be overcome, and for this reason they shall be carefully examined in Section III.4.

One of the major obstacles to the existence of programmes of this type is the inadequacy of current institutional structures that does not even permit some necessary actions to be considered. There are several main aspects leading to this inadequacy.

III.2.1. Distribution of decision-making authorities between different jurisdictional levels

In OECD countries the implementation of suitable programmes is made difficult by the overlapping of territorial jurisdictions. Each of these jurisdictions is assigned areas of decision, which furthermore differ from one country to another. Congestion is rarely limited to the areas covered by the territorial authorities who are empowered to deal with the problem.

For example, the San Francisco Bay Area consists of nine counties each of which is responsible for congestion management in its territory. In London, the 33 boroughs have decisive power as regards road improvements. In France, the different roads in an urban zone come under different territorial jurisdictions (the State, departments and communes).

It should be stressed that rational transport operation demands responsibility to be exercised at network rather than territorial level.

III.2.2. Multiple agencies with various responsibilities

The various territorial levels of decision have created administrations and agencies which are responsible for a particular traffic related task. Any congestion problem usually involves several agencies, but as a rule no agency has the task of tackling congestion as a whole and therefore no agency is in a position to deal with it effectively. In many cases the absence of a single authority hinders the implementation of comprehensive programmes.

For example, in several countries, an accident involving an international freight vehicle which takes place on an urban motorway brings into play the customs authorities if the goods are under customs control, the police, the motorway operator, the emergency services, etc. Under these circumstances it is difficult to assign priority to re-establishing traffic service levels.

Furthermore, the separation of responsibilities between Ministries may lead to potential conflicts between (and sometimes within) Ministries. In many countries, the powerful road construction administration is separated from that entrusted with traffic management. Hence the main response of this type of institutional structure to a problem of congestion tends to focus on capacity increase (which was the major need in the past) rather than manage congestion and demand.

III.2.3. Priorities and distribution of funds

The budgetary decision making which is necessary in order for any transport policy to be instituted is usually the victim of historical inflexibilities. Budgets are traditionally channelled towards the construction of new infrastructure, which constitute concrete projects (particularly important for voters) and the operation of existing roads receives only the bare minimum. It is difficult to re-channel budgets because of the underlying administrative structures which support them.

Often, decisions to implement measures are taken not in function of their own efficiency, but according to the credits and subsidies that may be assigned to these measures by the State. For instance, in Germany, municipalities will implement bus lanes only if traffic signal preemption is added to the project, because it is a way of getting federal funds. In France, a lot of measures concerning traffic management may be implemented because they are labelled "road safety". Subsidies from the European Community may be of decisive importance for implementing some projects.

III.2.4. Inadequate participation of the public

The implementation of transport policies suffers from a failure to communicate with or listen to the public. Even at the stage of the public enquiry which precedes the construction of new infrastructure this failure generates poorly controlled opposition movements. However, the need for communication is even greater in the context of road operation. For the purposes of congestion management it is essential to gain the understanding of the public as this is necessary for the effectiveness of many measures. For example, the SIRIUS real-time road user information system on motorways in the Ile-de-France region requires a high degree of communication to- and feedback from the public. Road-vehicle communication systems currently under development demand the participation of road users and must therefore provide them with a service.

However, the existing institutional systems have difficulty in meeting these needs for interaction.

III.2.5. Orientation towards short term planning and quick results

The aim of this study is to describe congestion management measures which can be introduced in the short term. However, while it is possible for immediate actions to relieve congestion, comprehensive congestion management programmes can only be envisaged in the context of a long-term strategy.

Such a strategy is difficult to define and implement for several reasons:

♦ Decision makers are not always very involved in defining policies to be applied after the end of their electoral mandate,

♦ Efforts to enforce implemented measures are not always carried on after a few years,

♦ The different sets of measures taken by successive elected boards are not always consistent.

In some countries, effective attempts to overcome these difficulties have been made. In the United States, the new 1991 Federal Intermodal Surface Transportation Efficiency Act should, if budgetary mechanisms are modified, make it possible to introduce congestion management programmes in the long term. In the Netherlands, the second Transport Structure Plan sets the objectives for the 2010 horizon (as a probability of less than 2 per cent of being subjected to congestion) before specific congestion attenuation programmes have been either developed or implemented.

III.2.6. Insufficient evaluation of congestion management measures

The development of congestion management programmes suffers from the lack of evaluation of transport strategies. This evaluation must be carried out to find out the effectiveness of the different measures and the interactions which occur between them, thereby making it possible to develop a programme which enables the goals to be achieved.

The individual measures described in Chapter II have not always been evaluated, and, frequently, the evaluations which are undertaken are incomplete or inadequate. To an even greater extent, the programmes which have been introduced have not undergone thorough evaluation.

The reason for this omission lies, obviously, in the difficulty of carrying out evaluations of this type.

Firstly, data may not exist, so that evaluation is not possible. Then, the different measures have multiple effects, many of which are still poorly understood. This applies, for example, to the effects on the economic environment (transformation and localisation of activities), effects on social structures (rich and poor neighbourhoods), effects on the natural and urban environment (pollutant emission, severance effects), effects on the competitiveness of towns and cities in the context of the increasing integration of economies, particularly in Europe.

The transport system is a complex system which encompasses a large number of players, each with his/her own approach and legal status, and without a strict hierarchy (as has been seen above). The system also involves different modes and combinations of modes, and interacts with the housing market and the organization of society. It tends to be structured as a network level, the interdependencies of which are still unclear, which results in an increase in both its performance and vulnerability.

Another problem may be the reluctance to conduct and publish analyses which would reveal too clearly the impacts of any course of action; administrations may prefer to respond to public concern by instituting measures which at least appear to address the problem, even though there may be doubt as to their effectiveness. Nevertheless, in some cases, significant efforts have been made to produce accurate evaluations. For instance, the european DRIVE programme is refining a detailed evaluation procedure for individual ATT applications, but few numerical results can be expected before 1995.

III.3. KEY FACTORS OF EFFECTIVE PROGRAMMES

To overcome the difficulties that have been described above, it is important to take into account different factors that may lead to the success of a congestion management programme. These factors are of different nature, that may be conveniently presented in a hierarchy. At the bottom is the basis for implementation of programmes, that is the knowledge of the way the traffic and transport system works, the technical aspects involved in measures, which we can call the planning and project development process. Just above, are the financial and economic factors that are important for the profitability of projects and the knowledge of demand. Then follow organisational and institutional factors that are needed to implement programmes. Above them, legal aspects and regulations may be involved. Taking into account all these factors, the decision making process has the role of conducting programmes. At last, the higher level is of course the one of society as a whole, that is sovereign in our democratic countries.

The present section will discuss the importance of those factors, by referring to the examples of congestion management programmes presented in Section III.4.

III.3.1. Planning and project development factors

Systematic planning is an essential ingredient in the development of effective congestion management programmes.

First, a good knowledge of the transport phenomenon is needed. In this respect, it may be certain that congestion, as a problem, cannot be completely solved; there is an equilibrium between supply and demand for a particular level of congestion; the purpose of the trip-making policy being to fix the parameters of the point of equilibrium: the problem can only be managed. For that, planning helps to define a course of action by identifying the systems conditions and the marketplace in which measures are needed.

In this regard, the examples primarily have demonstrated the need to apply a combination of measures to remedy congestion. Thus, examination of the Amsterdam-Utrecht corridor (See Section III.4.3) has shown the potential benefits of combined policies which couple a limited increase in highway capacity with an increase in rail capacity and public transport services.

Most of the time the utilisation of modes other than passenger cars or trucks is fundamental, as in Switzerland with the development of combined transport or in the Netherlands with policies to encourage carpools, bicycles and public transport. In fact, in congested urban areas, the flow of road traffic is essentially determined by the quality of service of mass transport modes, as has been stated by Mogridge. However, many studies show that the different modes are not perfectly interchangeable, as is the case with the Amsterdam-Almere link, where the introduction of a rail service has not altered the number of motorists, or their behaviour, as the train has mostly taken the place of buses.

In order to significantly reduce the level of motorised traffic it is necessary to combine encouragement (pull actions), such as the development of alternative modes, with dissuasion (push actions) such as restrictions on car use. The Groningen transport plan is a perfect example of this (See Section III.4.4).

Second, a sound evaluation of alternative strategies is to be made prior to implementation of any measure. Effective tools for that are the study of effects, cost-benefit analysis, multi-criteria decision techniques. The systematic implementation of such tools ensures a correct ranking of different alternative policies and enables the choice of the most efficient one. Such studies must be based on accurate data. For that, one should identify what data is needed, why and for whom it is needed, and how it is to be collected. These data enable project analysis and evaluation in order to effectively understand the impact of the congestion management measures on such things as mode choice, traffic volumes, and traveller behaviour.

Third, the monitoring and evaluation of the achievements of the project undertaken must be made to ensure that the congestion management measures implemented are achieving the results that are desired. Collecting information for project monitoring and evaluation needs to be considered as standard operating procedure when planning and implementing congestion management measures.

An example of this type of planning in the United States is the Freeway Management Programme for Interstate 5 in Seattle, Washington. On this project, a package of congestion management measures was planned on the following objectives: 1. to improve freeway operation, 2. to reduce congestion and congestion-related accidents, and 3. to maximise the people-moving capacity of the freeway through ramp metering and preferential lanes for buses and carpools. Evaluation data is collected to monitor how effectively the measure has achieved the goals and objectives set forth.

Finally, another important feature of successful programmes is a gradual implementation: often, programmes may begin with small scale demonstration projects, and, if they prove to be efficient, they may be extended to a whole area. Nevertheless, in certain cases, measures may not be efficient at a small scale because of side effects (for instance, high parking fees leading to diversion to neighbouring areas), but may be a very efficient way of managing congestion in a whole city or corridor.

III.3.2. Financial and economic factors

Although financial and economic factors are directly linked with the planning process, we can stress here the following points.

Funding sources for congestion management should be identified and allocated, at different levels of jurisdictions. It is certainly beneficial to have a dedicated funding source for congestion management, in order to implement effective measures.

The extension of tolling and road-use pricing is spreading, at least conceptually, and will have a vital part to play in achieving an acceptable and social equilibrium:

- Toll roads, bridges and tunnels have existed for a long time, as soon as the nineteenth century in the United States. In Italy, France, and Spain, the motorway network has been developed thanks to toll collection, and extensions are still planned. Other countries, where motorways are generally free of charge, debate about building new toll roads, as Britain or Germany. For Central and Eastern European Countries, tolling is an important means to finance a motorway network.

- Theoretical developments about road pricing appeared in the 1960's. Application began in Singapore in 1975. Discussions about application in the United States and in Europe have been speeded up since the late 1980's, but has not yet led to real congestion pricing: toll rings in Norway are only a means of raising funds for building new infrastructure.

The sharpest and easiest to apply of measures to attack directly the level of demand on road space is the pricing and control for parking, which requires sensitive treatment according to the many individual characteristics of the area concerned.

Involvement of the private sector is increasing and will be more and more needed to solve traffic problems. As to planning, the private sector has a role to play and private consultants may make less biased studies than administrations in charge of the project. Also, for a considerable time the private sector has been involved in constructing and equipping roads, and even in finance and operation in the case of privately operated roads. With current budgetary restrictions this type of road is likely to increase in number, particularly in urban areas.

In the light of the development of new technologies -- IVHS in the United States, ATT in Europe, ARTS and UTMS in Japan -- the private sector is starting to play a key role in developing operating techniques, in standardisation processes, and in providing services to users. Particular mention can be made of the involvement of telephone companies and the wide distribution of radio telephones.

III.3.3. Organisational and institutional factors

The congestion management programmes which have been applied, or even more, those which have not been applied in OECD countries, have shown the need to co-ordinate the role of the different parties involved in transport and traffic. For this purpose, **co-ordination mechanisms** must be developed:

- between different public agencies, at national, regional or local level,
- between urban and inter-urban zones,
- between the bodies responsible for constructing, operating and monitoring highways, and especially with traffic management and surveillance authorities, including the police,
- between road users, freight and passenger road transport operators, residents associations, etc., and the bodies which have already been mentioned above,
- with the operators of other modes of transport.

Beforehand, it would also be essential to build up adequate collaboration with land use planners. The fundamental problems which arise are the creation of a hierarchy for decisions and the assigning of responsibilities and the resources to carry them out.

Furthermore, all the measures which are necessary in the context of international political strategies to facilitate transnational movements must be introduced, by achieving harmonisation and standardisation at international level.

The generalised introduction of measures which are currently implemented on a local basis would improve effectiveness and also overcome specific opposition from minority social groups who sometimes feel that their rights have been encroached upon.

Beyond the simple function of co-ordinating the role of existing players, implementing comprehensive congestion management programmes demands the defining of responsibilities. These are often of new types which are required to ensure the correct operation of the programmes. It is often necessary to create new units to carry out these functions, without which the distribution of functions between existing departments is in danger of paralysing the creation of programmes. These new units may be a separate organisation, a task force, or a working group made up of representatives from agencies with responsibilities for the project.

In certain situations it may even be beneficial to create an authority to organise transport and traffic at metropolitan level, for example, in order to overcome the traditional divisions between different territorial responsibilities and different modes of transport.

Thus, in London the post of Traffic Director has been created, with authority over the various boroughs for the purpose of implementing the Red Route scheme (See Section III.4.2).

In the United States, special task forces are regularly established to coordinate the implementation and operation of congestion management measures, especially during the reconstruction of major motorway segments in urban areas. For example, task forces to plan and monitor congestion management measures are used in Pittsburgh, Pennsylvania (See Section III.4.8); Chicago, Illinois; Boston, Massachussets; Houston, Texas; San Francisco, California, and Seattle, Washington. TRANSCOM, in the New York, New Jersey area is a separate agency designed to create coordination among the numerous agencies responsible for traffic management. The National Incident Management Coalition is an attempt to coordinate efforts in reducing the impact of incidents on traffic congestion.

III.3.4. Legal and regulatory factors

New regulations are often needed to implement congestion management measures.

In the United States, congestion management is becoming part of Federal, State, and local transportation regulations. These regulations will ensure that congestion management measures become an integral part of transport policy at all levels of government. The Federal and State environmental regulations will also require the involvement of the employers, and give them responsibilities for ensuring that their employees get to work in the most efficient mode available.

Occasionally, the introduction of congestion management programmes cannot be achieved without modifications to the law. Sometimes, the law needs to be modified to allow new measures (this was the case in Spain, for example, for the introduction of parking charges on the road network); sometimes, the law determines the aims to be achieved and the rules to apply (this is the case in the United States for the state congestion management programs).

In the United States, the new Congestion Management System Requirements in the 1991 Intermodal Surface Transportation Efficiency Act (ISTEA), is innovative in that it ties Federal transportation financial aid to the development of a congestion management programme. That is, congestion management is a major part of the U.S. transportation policy and in order to get funds from the U.S. Department of Transportation, the state and local area must also have a policy and a programme for congestion management in place by 1995.

III.3.5. Policy development factors

In practice, the technical or institutional difficulties associated with the introduction of congestion management programmes may be fully overcome on condition that there is a clear transport policy. A lack of precise political objectives often constitutes a severe handicap to the introduction of coherent programmes and is likely to result in resources being sprinkled about (a little for roads, a little for public transport), which fails to cure the basic congestion problem.

The need for a transport policy applies at national or regional level, of course, but is of greatest importance at town or city level, which is where most congestion problems occur. As has been seen above, in the United States, state and local transportation policies are forced by law to include congestion management. However, elected representatives, particularly at local level, when trying to make a genuine coherent comprehensive policy, have to face the influence of pressure groups with conflicting opinions and concerns. A strong political leadership is vital as any effective policy will encounter intense opposition from many individuals.

Moreover, it is important for policy-makers to understand that long-term success in the management of congestion demands relies not only on the pronouncement of a policy, but also on persistence in its implementation and enforcement in the face of continuing pressures to defeat it.

A deliberate transport policy enables a variety of fundamental decision making processes to take place. Budgetary decisions are of course amongst these -- which projects shall receive public funding? Decisions are also to be made regarding the sharing of road space -- how should public space (which is in short supply in urban areas) be shared between the various categories of road users, delivery vehicles, public transport, taxis, bicycles, pedestrians, parks, markets, use of footpaths by cafes, parking, and traffic? These decisions are the result of compromise between different conflicting requirements. Underlying them is a genuine choice regarding life style and the urban environment.

In a democratic country this policy issue involves discussion between citizens and is not merely a subject of debate between technical experts. Their role is limited to illuminating the key technical and economic factors.

III.3.6. Social factors

The citizens debate should, of course, take into account the opinion of road users and the public, as it is ultimately for these groups that congestion management programmes will be implemented.

Particular attention must therefore be paid to the following points:

◆ The acceptability of measures should be a prior condition of the implementation of programmes. For example, the Dutch government abandoned the Rekening Rijden road pricing project because of the lack of political courage due to the fear that the public would oppose it.

♦ Interaction with the public should be permanent. In particular, continuing dialogue and mutual understanding are needed between technical experts and the lay public.

♦ The measures must be equitable, both spatially - not disadvantaging one area with respect to another by restricting access to a town centre, for example - and socially - not reducing the travel possibilities of any social class, thus if tolls are introduced for motorised traffic, compensatory measures should be introduced to improve other modes.

Usually, the citizen will have a negative reaction when measures are to be taken to restrict, or make more expensive, the use of his/her car, but in the same time he/she will support the idea that a decrease of car use and an increase of other modes, especially in cities, would be beneficial. Research in the European Community Countries has shown that the majority supports investments on public transport and bicycling at the expense of investments for private car transport: 80 per cent think developing public transport is efficient to solve traffic congestion problems.

Congestion management programmes will have the greatest chance of success in mastering congestion and making it compatible with the lives of all citizens, if transport and traffic policies are given the support of society.

III.4. EXAMPLES OF GOOD PRACTICE

To illustrate the factors contributing to effective congestion management policy, the OECD study group has chosen to present 11 examples in different countries and in different contexts (city, regional, state or national levels). Table III.1 shows the most important aspects of each example.

III.4.1. Integrated transport planning in Zurich, Switzerland

Objectives and principles

Zurich is the main town (365,000 inhabitants) in a region with about a million inhabitants. After the approval of the "people's initiative" for projects to speed up public transport in 1977 (SFr 200 million), resources were provided to implement speed-up measures and, more significantly, a majority of the population concerned expressly agreed with a transport policy giving priority to trams and buses. In 1979, a directive issued by the Town Council instructed the municipal authorities to give precedence to public transport in any conflict involving the various interests in the transport sector. In 1981, the voters in the Zurich canton approved the regional light rail project. In 1987, the Town Council consolidated its plan with the aim of promoting public transport and reducing motor vehicle traffic.

Measures and application

The speed-up programme focused on three objectives: keep ways unobstructed by cars by creating dedicated tracks and separate bus tracks, give maximum priority for public transport vehicles at traffic lights directly operated by trams and buses, monitor the public transport network by a computerised control system with automatic location data acquisition.

Table III.1. Factors contributing to effective congestion management policy in different examples

Factors/Examples	Planning and Project Development	Financial and Economic	Organisational and Institutional	Legal and Regulatory	Policy Development	Social
Integrated Transport Planning in Zurich, Switzerland	XX	X	X		X	X
The Priority Route Scheme (Red Routes) in London, UK			XX	X		
Amsterdam-Utrecht Corridor Study, the Netherlands	XX	X	X			
Regional Transport Plan in Groningen, the Netherlands	XX	X	X			
Traffic Demand Management by Companies in the Netherlands	X	X	X		XX	
Urban Travel Plans in France	(X)		(X)		XX	
Road Operation Master Plan in France			XX		X	X
Highway Reconstruction Project in Pittsburgh, PA, USA	XX	X	X			(X)
Public/Private Agreements for Congestion Management in Montgomery County, MD, USA	X	(X)		XX	X	
California's Congestion Management Programme and Washington State's Commute Trip Reduction Law, USA	X	(X)		XX	X	
The National Congestion Management System, USA	XX	X		X	X	
The Swedish National Committee on Urban Traffic	X	XX	X	X	X	
Area licensing scheme in Singapore	X	XX	X	X	X	

(X)   Secondary factor
x   Significant factor
XX   Most important factor.

The regional light rail system went into service in 1990; it covers the entire canton of Zurich and some neighbouring municipalities. The main part of the SFr 2 billion regional light rail project was the extension of the old central station in Zurich with the construction of a through station and the Zurich mountain tunnel. The project also included the creation of new connections, diametrical lines, a basic interval timetable, a tariff agreement and the incorporation of the individual transport operators within the Zurich Transport Authority (ZVV).

In the mid-1980's, transport operators worked out the Zurich 1990 Plan: a connection with the regional light rail for all inhabitants of the town, new services for areas with poor communications, market-oriented services (more room and shorter intervals by means of denser services).

Effects

With the introduction of the regional light rail system, seating capacity in the morning peak has been raised on average by 20 per cent. The number of passengers rose by 20 per cent in the first year of operation and by a further 5 per cent in the second year.

From 1985 to 1990, 30 per cent more travellers were carried by municipal operators; with around 470 public transport trips per inhabitant per year on municipal transport services, Zurich holds the world record.

Despite these results, there has been no perceptible effect on the volume of car traffic in the town of Zurich; the traffic flow at the town boundaries in 1991 was the same as in 1990.

The car is attractive for the individual trips, even when the quality of public transport is very high. In addition to encouraging public transport, it is essential to limit the attraction of private car travel.

Institutional problems

After more than 20 years of intense activity, the transport policy has for a long time shaped the thinking and action of all those working in local government. Pragmatically, advancing by smaller or larger steps, they have assembled the many components to form a whole. For example, the decisive factor in achieving the highest standard of public transport is to be found not so much in the major individual projects as in the sum of all the steps taken, particularly the small ones.

A regional approach should be taken to municipal transport problems and the implementation of measures. Even if there is growing understanding for Zurich's attitude in districts around the town, possibilities of conflict still exist.

III.4.2. The priority route scheme (Red Routes) in London, U.K.

Objectives and principles

Along a priority route, the management of traffic is reviewed in detail. The space previously occupied by illegally parked vehicles is then reallocated to buses, pedestrians, cyclists, essential traffic such as delivery and emergency vehicles and provision for legal parking and loading. The concept was originally intended to unlock existing strategic traffic routes within London and parallels were drawn with unclogging the Capital's arteries. This led to the term Red Routes being commonly used.

Measures and application

The Secretary of State tested the various measures which are likely to be used on Red Routes on a 12.5 kilometre Pilot Scheme in North and East London:

♦   620 new free short-term parking spaces were introduced,

97

- 1200 stopping places were introduced,
- 132 entry treatments were built at side road junctions,
- 4 new cycle crossing facilities were provided,
- 17 new improved pedestrian crossing facilities were provided,
- existing 2.9 kilometres of bus lanes were increased to 5.02 kilometres,

Effects and cost-effectiveness

A summary of some aspects of the Pilot Scheme follows:

- Bus journey times were reduced by 8 minutes to almost 29 minutes and a 33 per cent improvement in reliability was achieved,

- 1700 more passengers each week (up 3%) travelled on bus route 43 compared with a 1 per cent decline in London generally,

- 11 per cent increased flow on A1: +9 per cent from diverted trips, +2 per cent unaccounted for,

- Journey times were down 6 minutes to almost 20 minutes and there was a 40 per cent improvement in reliability,

- Casualties were down by about 200 in the first 18 months (a reduction of 17%).

Costs	
Design and works	£4.5m
Enforcement	£0.9m/year

Benefits	
Saving casualties	£4m/year
Journey times/Savings	£4.5m/year

Institutional problems

The development of Priority Routes needed a Road Traffic Act and the establishment of a post of Traffic Director for London in 1991. The principal duties are to develop and coordinate the introduction of Priority Routes and monitor their operation, many of which lie on local authority roads. This extends to having control over any measure that is proposed on any other road which would affect the operation of these routes including traffic signal timings etc.

The Red Route programme started in 1992 and its introduction will be completed in 1997, by which time there will be 500 km of Red Routes.

III.4.3. Amsterdam-Utrecht Corridor Study, The Netherlands

Objectives, principles and institutional problems

The corridor approach includes the optimisation of the supply of infrastructure by applying the translation of the national limits for the growth of automobility to a regional scale and predicting the effects of policy measures (national as well as regional). The possibilities for a change in the modal split will be studied.

The second main point is the "open planning process". In the Netherlands, the physical planning is decentralised to the local authorities. But the groups that may influence the decision of the local authorities are also involved in the study; these include the chamber of commerce and in some cases even pressure groups.

Measures and application

Starting in 1990, the corridor approach is applied to the corridor Amsterdam-Utrecht.

The city of Amsterdam is an important centre of trade and financial business, while the city of Utrecht is a centre of logistic and computer services. The two cities complement each other, thereby creating a lot of traffic vice versa, commuter traffic as well as business traffic.

The objectives are to take a project decision (number of highway lanes, number of railway tracks, way of construction) and to create input for policy measures (parking, transportation, demand management, car pooling).

Four alternatives are under consideration:

♦ The zero-alternative represents the existing infrastructure and the existing policy combined with the traffic demand in 2010. Of course, this alternative results in dramatic congestion on the highway and in bulging train; it's not a real alternative;

♦ In the zero-plus alternative, a mix of policy measures is introduced, based on the Transport Structure Plan. This is combined with the expansion of railway; the highway is still full of congestion;

Alternatives

	Highway lanes	Railway tracks	Policy-mix
0 Alternative	3	2	none
0+ Alternative	3	4	N°1
B1 Alternative	4	4	N°1
B2 Alternative	4/5	4	N°2

- The alternatives B-1 and B-2 combine the expansion of the railway system with a limited expansion of the highway. The difference between these two alternatives is the different mix policies. Compared to policymix 1, policymix 2 has a less restrictive parking policy and a lesser price index for car costs.

Effects and cost-effectiveness

There is only a minor change in mobility between the zero-plus and the B1 alternatives; this may indicate that, in this corridor and with this mix of policy measures, the existence of congestion does not influence car use any more.

The difference between the B1 and the B2 alternatives shows the importance of policy measures and the effects that these measures may have on the demand for infrastructure; this is clearly shown by the extra highway lane that is needed to absorb the amount of traffic.

If we compare the figures of this study with the national figures, we find that the demand for transportation by car in this corridor is less than in the country as a whole.

There is an increase of passengers on trains. Partly, this is because of the policy measures, which cause some car drivers to become train passengers, but for the greater part, this change is caused by new passengers on the train, as a result of the upgrading of the railway system.

For the transport of goods, the most extreme scenario does not permit the use of lorries over more than 150 km. Even in this case, a reduction of only 12 per cent of the number of trucks on the highway is achieved.

Evolution of the regional index of mobility compared to the existing situation (level 100)

	Car trip index	Train and bus trip index
Now	100	100
0 Alternative (2010)	155	178
0+ Alternative (2010)	121	232
B1 Alternative (2010)	121	234
B2 Alternative (2010)	137	215

III.4.4. <u>Regional transport plan in Groningen, The Netherlands</u>

Objectives and principles and institutional problems

In the early 1980's, a number of cities had a traffic circulation plan, as a result of the national transport policy concerning cities with a municipal public transport company. Groningen is one of these cities.

In the city of Groningen, capital of the province of Groningen, the plan was aimed at changing the traffic circulation drastically. At the end of the 1980's, the changing position of the city of Groningen to the centre of the northern region of the Netherlands gave new impulses to a wider initiative: the development of the transport region of Groningen.

There is a cooperation of the national government, 2 provinces, 26 municipalities, 1 municipal bus company, 3 regional bus companies, the National Railways and the Chamber of Commerce. The purpose of this cooperation is to prevent that a too strong growth of car traffic in this district would result in problems concerning accessibility, environment and livability. Three levels of management work together in a vertical way (government, province and municipality) and horizontally, provinces and also municipalities work together. This cooperation crosses borders of provinces. Furthermore, it is a public-private partnership, on a voluntary basis.

In January 1991, this cooperation led to a covenant, which was signed by all partners. It consists of a policy programme and a long range programme of execution.

The strategy consists of stimulating alternative modes of transport, not stimulating car use (and even, where it is possible, discouraging it), and other supporting measures.

These are called the three P's: Pull, Push and Publicity. There is a need for a balance between pull and push measures, with a special attention for road safety and transport of goods.

All the measures must be done in close connection, leading to a mix of measures.

An adequate town and regional planning is a very important instrument in the programme; the attempt is being made to integrate public transport and bicycles. Living, working and facilities are situated in such a way that car use cannot be stimulated. New residential areas and offices will be situated near junctions or stops of public transport. Parking facilities for commuters are limited. Within the town, residential areas, offices and facilities are constructed within cycling distance. For existing situations, the regional and town planning is adjusted as much as possible, so that public transport and bicycle facilities are being improved.

Measures and application

Measures for bicycles are: cycleway network (radial and tangential connections), special facilities (cycleways or lanes, differently coloured asphalt, two-way traffic for bicycles in one-way streets, special lanes near traffic lights, special short-cuts to the right to avoid red lights, bicycle stands and grips, guarded shelters...).

Measures for public transport include: integration of various kinds of public transport (train for long distances, bus for medium-long distances, bus or bicycle for short distances), bicycle sheds near stations and bus stops, short bicycle and pedestrian routes to facilities, traffic signal pre-emption for buses, bus lanes, special shuttle services for special events.

Measures for ride-sharing are: ride-sharing parks in strategic places, ride-sharing mediations organized through many companies and 3 municipalities, possible future use of bus lanes.

Moreover, a strong controlling parking policy applies to the town: park-and-ride lots, car parks near entrances of the town centres, parking meters near the centre with high tariffs for short time.

Costs and effects

All the proposed measures have been laid down as a mix of measures in a long range programme of execution 1990-2000.

This amounts to a total sum of 220 millions guilders. Up to now, 20 to 25 million guilders is invested on an annual basis, of which the contribution of the national government is 12-15 million guilders. The rest is paid by the regional partners. Two-thirds of the amount are invested in the town of Groningen.

A programme of evaluation is being developed. However, a number of indications of effects can be given from now on:

♦ problems at a number of traffic hold-ups have been taken away,
♦ the new facilities for bicycles have been used well,
♦ the approach concerning a number of unsafe spots has led to a decrease of the local accident rates,
♦ the constructed car pool places have been used well,
♦ companies actively cooperate in projects of ridesharing mediation,
♦ the local shopkeepers in the town of Groningen are in favour of extending the pedestrian areas,
♦ nuisance caused by cars looking for a parking place decreased because of "full/free" signalling.

III.4.5. Traffic demand management by companies in The Netherlands

Objectives and principles

Since 1989, Traffic Demand Management by Companies (TDMC) has been used to reduce car use by cutting back commuting travel. Traffic Demand Management by Companies is stimulated by the Dutch government: at 7 sites and in 25 companies, experiments were (partially) financed. The Dutch Ministry of Transport and Public Works supports the experiments, develops instruments and will produce an overall evaluation.

Measures and application

Traffic Demand Management by Companies is applied all over the Netherlands; it is a part of the National transport policy to reduce the growth of car traffic (the second transport structure plan). Most of the experiments have started; for some projects, an evaluation study has been carried out.

Traffic Demand Management by Companies is implemented in sites such as: Schiphol Airport, The Hague, Amsterdam, Utrecht..., with companies such as: the Dutch Ministry of Transport and Public Works, the Traffic Engineering Division and Regional Board South-Holland, the Province of Groningen, Nedcar (Volvo)...

Examples of measures are: improving public transport, stimulating car pooling, parking measures (restrictions), company transport, communication, marketing and public relations activities, change of travelling allowances, bicycles facilities...

Evaluation

Recently, an evaluation study among 17 companies has been carried out. Some factors appear to be critical for success:

A comprehensive approach is the first factor, which means that both stimulating measures for the alternative modes and measures discouraging single car use need to be taken simultaneously;

The location of the work site is the second factor. Restricting the parking area at the work site may not be effective when too many public parking places are available near the site: car driving employees can easily find a place outside. The same holds for raising parking fees. For this type of measures, co-operation with the municipality is necessary to tune policy. Moreover, locations in the centre of the cities often provide better accessibility by public transport and bicycle, than locations in the outskirts of the towns. For the latter, the chances of alternative modes are smaller.

The average travel distance to the work site is the third factor. When companies or institutes have been located for a long period of time at the same place, employees generally live closer to the work-site, than employees from companies at relatively new sites. In the first case, less employees commute by car. Moreover those employees who still do, appear to be more easily moved to other modes than the employees of companies at relatively new sites. Since travel distances are short, the bicycle is generally favourite here.

The courage of the employer is the fourth factor. Most people are not easily prepared to change habits. This is also true for travel habits and for employees. Changing the financial conditions for commuting is one way to influence travel behavior. Those conditions are generally not sufficiently changed by just upgrading the alternatives for single car use. This is often rather costly and eventually does not affect the car driver. Redistributing the available funds for allowances, resulting in better travel conditions for the alternative modes at the expense of those for the single car driver is more effective. The evaluation among 17 companies and institutes has shown that generally the transport plans with the lowest costs have resulted in the largest reduction of car use (up to 30 per cent of car-kilometres). In many of these cases, the low level of costs has been reached by redistributing allowances. To introduce drastic changes in travel conditions, a certain courage from the employer is required. It is not an easy job to sell worsening commuting conditions to the majority of one's employees. Often these changes can just be realised after a period of a gradual shift in thinking about the subject, and when most of the employees agree with the reasoning behind the measures.

Institutional issues

The Ministry of Transport (Regional Board), local authorities, companies, Chambers of Commerce, industrial insurance boards, consultants are some of those involved with Traffic Demand Management by Companies. Some of the Traffic Demand Management by Companies organizations are similar, there are often differences (project management, some Traffic Management associations are private companies). A project team of the Ministry of Transport has a coordination role.

III.4.6. Urban travel plans in France

Objectives and principles

The Internal Transport Orientation Law (1982) defined the urban travel plans: they must present the general principles of the policy of transport, traffic and parking organization in the urban transport area. They aim at a more rational use of cars and they ensure the right insertion of pedestrians, cycles and public transport. They are accompanied by a study of their financing modes and of the operation costs of included measures. They are subject to a public survey. They must be approved by the competent local authority of the concerned geographical area.

The plan contents the terms of the local travel organization policy and the presentation of a succession of actions and operations. The plan must define the structure which will be responsible with its application.

Measures and application

About 40 towns have studied an urban travel plan, like Annecy, Bourges, Grenoble, Lorient, Montpellier, Nantes. Four types of aims have been brought up: to improve the external productivity of public transport (bus lanes, right-of-way at traffic lights, park-and-ride lots), to give a new value to forgotten modes of transport (pedestrians, cycles), to establish a reference framework to make the political choice easier, to develop user and inhabitant information.

In the short-term, urban amenities have been provided, just as a restructuring of public transport network, fare modifications, implementation of dialogue structures. In the medium-term, measures concern the urban organisation, the infrastructure master plan.

In Grenoble, for instance, the main chosen options concern the urban high speed roads (opening of the south by-pass), the creation of a tramway line, a study of a second tramway line utility, the general improvement of the public transport network.

Institutional problems

The responsibility of the elaboration of the urban travel plan belongs to the local transport authority.

In a lot of cases, some difficulties exist at this level: the traditional local structure is municipality, but the spreading of urbanization lead to the constitution of built up zones covering several municipality areas, often without having any global authority. In France, there is often an intercommunal authority for public transport, but, in many cases, there is no authority of this type to manage the general traffic problems and thus the ensuing congestion.

III.4.7. <u>Road operation master plan in France</u>

Objectives, principles and institutional problems

Numerous actions have already been undertaken by the different road services (local department services, motorways companies, local authority road services, police forces) to limit the consequences of the traffic problems and provide the users with information on the traffic conditions they will meet.

At the end of 1991, the Ministry of Transport decided to extend those actions to the whole national network in the frame of a road operation master plan (*Schéma Directeur d'Exploitation de la Route*), which will define for the different categories of roads of the network, the operation level aims, the organisation to be set and the corresponding means.

Concurrently with this first technical approach, a dialogue has been started with other concerned ministries (the Ministry of Interior, the Ministry of Budget, the Ministry of Defence...) and the local authorities, aiming at clarifying the rules governing the attributions and the jurisdictions of the road operation services, to allow the integration of the whole networks in a consistent and global operation policy.

Measures and application

This plan includes: a network classification and propositions for new operation structures, if needed. The road operation includes three main missions: road-practicability maintenance, traffic management, travelling assistance.

The road-practicability maintenance comprises the whole on site interventions, meant to maintain or restore, in case of perturbation, the using conditions of the road, the closest to the normal situation. It includes: the organization of the predictable interventions (works under traffic, convoys or manifestations) and the management of random perturbations and emergency interventions.

The traffic management comprises the whole dispositions meant to distribute the traffic flows in time and space, to avoid the appearance of uncertain or recurrent perturbations. It includes: preventive actions, needing a perturbation predicting capacity, implementation of alternative routes, information to the users; perturbation treatment, requiring elaboration of a traffic management plan, a permanent system for measuring the traffic flow conditions, operational intervention teams, real time road information means; linked networks control.

This management requires, in addition to the perturbation treatment appliances: the elaboration of a permanent strategy (networks hierarchy, choice of the priority flows...); a centralized and automated system for detection and processing of traffic data; automatic traffic control measure, by dynamic signalization of congestions and advised diversion (on expressways), or by traffic lights coordination (on normal urban roads).

The travelling assistance comprises the whole dispositions meant to improve the comfort and the safety of the road users, by spreading information on the traffic conditions. It includes: estimated information and dynamic information; the last require means for spreading in real time (variable message signs, dedicated radios or vehicle in-board equipments).

The plan defines 4 different road operation levels:

- ◆ level 1: expressways of large built up areas,
- ◆ level 2: heavily trafficked motorway corridors,
- ◆ level 3: moderately trafficked structuring axes,
- ◆ level 4: the remaining national road network.

A permanent dialogue becomes necessary between the different competent authorities (State, local authorities, networks managers, police authorities and forces).

All the questions are being investigated in the frame of a large concertation organised around the road operation master plan. The definition of this plan will be finished in 1993.

III.4.8. Highway reconstruction project in Pittsburgh, PA, USA

Objectives and principles

In order to overcome the complex coordination and political problems created by reconstructing a congested urban freeway that goes through many jurisdictions, the Pennsylvania Department of Transportation established a task force to guide the work of rebuilding the Parkway East (I-376) in Pittsburgh. The task force was made up of Federal, state, and local transportation professionals and decision-makers. Their charge was to develop and coordinate projects that would move people and vehicles safely and efficiently through the corridor during the reconstruction period.

The task force established a plan, called the "Pittsburgh Experiment" which revolutionised the concept of congestion management during major urban highway reconstruction. Because the plan was developed by upper management from all levels of government, decisions on how best coordinate and integrate activities were addressed and made quickly.

The actions taken by this task force were designed to do two important activities. First, the action were designed to safely control the traffic that needed to travel through the construction zone. Second, and more significantly, the actions taken were designed to provide people with alternative routes and modes so that they could make their trip without using the Parkway East.

Measures and application

The following actions were implemented by the task force:

♦ A significant public information and community liaison programme to let people know what delays will take place and what alternative are available;

♦ Improvement on alternative routes (such as traffic signal upgrades and roadway widening) to reduce any delay caused by the excess traffic and to make the roadways more attractive as alternates;

♦ Vanpooling to provide commuters, who don't live near public transit an alternative to driving alone to work;

♦ Improved bus service to make transit an attractive alternative;

♦ A new commuter rail service for those living near an existing railway freight line;

♦ Many new park-and-ride lots were built to provide access and intermodal transfer facilities to public transit, commuter rail, vanpools, and carpools;

♦ Special high occupancy vehicles ramps were established on the ParkWay East to allow only carpools, vanpools, and buses access to it during the reconstruction.

Effects

Because of the actions and coordination of the task force, vehicle volumes travelling on the Parkway East during the reconstruction were reduced by 60 per cent with travel delay being minimal.

Public acceptance of the project was high because the travelling public understood the problem and were provided with alternatives to driving alone.

III.4.9. Public/private agreements for congestion management in Montgomery county, MD, USA

Objectives and principles

Montgomery County is located immediately northwest of Washington, D.C., with a population of 700,000 and an employment base of over 250,000. The county has an "Adequate Public Facilities Ordinance" (APFO) dating back to 1973 that requires local officials to examine the adequacy of transportation facilities and services before approving new land developments. In recent years, the requirements have become progressively stronger and more effective, with increasing emphasis on public/private sector solutions. The ordinance has been the stimulus for a number of far-reaching traffic congestion alleviation requirements in the county.

Measures and application

The County's APFO requires that public facilities are adequate to handle new demands before development can take place. The ordinance technically covers all public facilities but it is transportation that has grown to be the most significant issue with respect to new development. If a proposed development will produce unacceptable levels of service conditions on the nearby roads, then public or private sector solutions must be found to either reduce trips or increase the capacity of roads.

The APFO spells out two approaches to determining the adequacy of public facilities to meet the demand of new development: (1) a local area review test and (2) a policy area review test.

The local area review is basically an intersection analysis on the roads in an area surrounding the site. For large sites (e.g. 1,000 dwelling units or 100,000 square meters of commercial space) this area could cover one kilometre or more. This review requires that traffic on the roads in the surrounding area operate at no worse than level of service E. If the required conditions for traffic flow are not met, the developer must work with the County to implement acceptable congestion management measures.

The policy area review follows the local area review and is intended to identify the downstream and upstream impacts of the additional development within one of the 19 general sub-areas of the County. Each of the sub-areas has predetermined levels of acceptable traffic congestion. The sub-areas with good levels of transit service have higher levels of acceptable traffic congestion than sub-areas with low levels of transit service. Based on these levels, a determination is made as to whether sufficient transportation capacity exits within the policy sub-area to allow additional development to take place.

Institutional and programmatic issues

To provide transport improvements to existing developments, Montgomery County has also embarked on an effort to establish transportation management organizations in certain sub-areas of the county with high density existing development. The organizations are voluntary, nonprofit, membership associations that includes both the private and public sectors. The organizations have the following primary purposes:

♦ To serve as a forum for private and public sector responses to transportation problems;

♦ To identify and implement actions to reduce traffic congestion, air pollution, and energy consumption;

♦ To facilitate orderly growth;

♦ To insure adequate access and internal mobility;

♦ To coordinate "demand management" programmes that include promoting transit and ridesharing services, spreading employee arrival and departure hours, and implementing parking management and other programmes to encourage the use of high occupancy vehicles for commuting;

♦ To organise, manage and promote bus or van programmes;

♦ To develop and coordinate common parking policies and other incentives and disincentives aimed at reducing the use of single-occupant cars;

♦ To help members meet trip reduction obligations;

♦ To conduct a cooperative planning programme to meet future needs.

III.4.10. California's Congestion Management Programme and Washington State's Commute Trip Reduction Law, USA

Objectives and principles

Much of the experience in the United States with comprehensive congestion management policies and programmes has been at the state level of government. The reasons for this are mainly that the States control a significant amount of public funds for traffic and transportation improvements for cities. By national mandate, states are responsible for environmental issues, such as air quality. States also play a very active role in local land-use and transportation planning activities and decision making. While traffic congestion is certainly a major reason for the policies or the programmes, it is the land-use and air quality issues that have been the driving force behind their political and legislative development and acceptance. In large measure, it is very appropriate for state government to take the lead in this endeavour.

The state of California has provided the most significant experience to-date with such programme. The California programme incorporates both supply-side and demand-side measures. The state of Washington also has a congestion management programme that emphasizes demand-side approaches more than supply-side measures.

The California Congestion Management programme was approved by public vote in 1990 and attempts to maintain acceptable levels of mobility and allow for the growth in employment centres. Counties and major cities are required to prepare congestion management programmes made up of at least five elements: traffic standards for major roadways, transit standards, travel demand management strategies, land-use analysis and a seven year capital improvement programme.

The congestion management programme requires that all streets, highways, intersections and transit systems meet some level of service standard within a seven year time horizon. The congestion management programme prepared by each city and/or county in the state must contain a seven year capital improvement programme that identifies the measures needed to maintain or improve transit and roadway standards. In addition, cities are requested to prepare deficiency plans that seek to mitigate existing and potential inadequacies in the jurisdiction.

The Washington State Commute Trip Reduction Law requires that cities and counties adopt commute reduction requirements for major employers. In accordance with the law, employers must implement travel demand management programmes to reduce the number of commute trips to the work-site. The following goals, based on a 1992 baseline of commute patterns for employers or cities, must be met by these employers:

A. A 15 per cent reduction in commute trips by 1995,
B. A 25 per cent reduction in commute trips by 1997, and
C. A 35 per cent reduction in commute trips by 1999.

The purpose of the law is effectively address congestion and air pollution problems, especially in the Seattle area. The law creates a partnership between the public sector (the city and county jurisdictions) and the private sector (the major employers) to deal with these problems. The law gets the employers working with the public sector to reduce the number of times and miles people commute alone to work. It calls on major employers, local governments and transit agencies to work together to achieve a reduction in commute vehicle trips.

Measures and application

Both the California Management Programme and the Washington State Commute Trip Reduction Law direct local (cities and counties) to focus attention on the congestion and air quality problems that they face. The congestion management measures generated by these requirements are generally applied at a particular employment site, an intersection, or along a transportation corridor. Often the measures are short-term or low-cost in nature.

In California, some examples of measures include traffic signal improvements, increased frequency of transit service, parking pricing programmes, and special lanes for high occupancy vehicles.

In Washington State, some examples of measures include express bus services, parking pricing programmes, flexible working hours, telecommuting, vanpooling and guaranteed ride home programmes.

Institutional and programmatic issues

The following issues have been most identified with these state programmes:

What level of coordination is needed between public agencies and between public and private sector agencies? What agencies or responsible individuals need to be involved in the coordination process?

What level of service standard is appropriate so that the goals of the programmes can be met effectively and responsively? Who needs to establish the level of service standards? Do the level of service standards need to be set low in order to show that the goals have been attained or high to show that the goals can be exceeded with some effort?

How will the state programmes and laws relate to Federal and/or local policies and programmes on congestion management?

How will the policies and programmes be carried out and enforced over the long term, especially for measures implemented by the private sector?

Effects

Both the California programme and the Washington state law are relatively new. The measures that have been implemented in both cases are primarily travel demand management oriented with some traffic signal improvements. No evaluation has been done that investigates any synergism between these types of measures.

Both state programmes have paid attention to travel demand management measures, primarily because of the emphasis on air quality problems. Transit improvement (like express bus service) and vanpooling have implemented to a large extend mainly because their effectiveness has been demonstrated. Both programmes are in the early stages of implementation, therefore it is too early to identify how effective the programme and the implemented measures are.

Attempting to solve transportation problems and improving the environment are very ambitious goals. Both state programmes identified following conclusions to-date in trying to achieve these goals through comprehensive congestion management policies and programmes.

- ♦ Congestion is not always bad and in fact encourages people to use transit, carpools a n d vanpools.

- ♦ Comprehensive measures are desired but have not been achieved, probably in order to show some success quickly.

- ♦ Congestion management policies and programmes seem to evolve into comprehensive policies and programmes. They generally do not start out that way.

- ♦ The development and implementation of a comprehensive congestion management policy or programme is generally costly and need to be supported with public funding in the early stages.

- ♦ The congestion management policies and programmes requires substantial coordination between public agencies and with the private sector. The coordination increases policy or programme effectiveness and cost efficiency.

- ♦ Comprehensive congestion management policies and programmes must address land--use impacts on transportation conditions.

- ♦ Comprehensive congestion management policies and programmes need to be reviewed and revised as needed on a continuous basis, in order to maintain its credibility and effectiveness.

III.4.11. The National congestion management system, USA

Objectives and principles

At the national government level, the Federal Highway Administration and the Federal Transit Administration of the United States Department of Transportation established, in 1993, regulations that will have each State government prepare a "Congestion Management System" (CMS). The goal of the

Congestion Management System is to identify and assess congestion and identify, evaluate and implement measures that make the most efficient use of the existing transportation facilities. The measures included in the Congestion Management System are supply and demand side. The requirement for the Congestion Management System was contained in the Intermodal Surface Transportation Efficiency Act of 1991. In addition to specific measures, the Congestion Management System will contain:

♦ Performance indicators to identify and monitor congestion, evaluate congestion reduction and mobility enhancement measures, and establish data requirements,

♦ An on-going data collection and system monitoring process,

♦ Congestion management measures and a process to evaluate those measures,

♦ Federal funding sources tied to the development of a comprehensive plan.

The Federal government is taking these actions because of its responsibility for national transportation planning requirements and because it is a source of funds to states for transportation improvements. The intent of the new Federal Congestion Management System requirements is to (1) ensure that congestion management measures from both the supply and demand side are incorporated into state and local transportation policies and programmes and (2) establish appropriate standard data needs so that the impact of congestion management improvements can be measured and reported to the united States Congress.

At the city and county levels of government, comprehensive congestion management policies and programmes have not generally been initiated and developed. Cities and counties that have these policies and programmes are required to do so because of a state mandated congestion management policy or programme. Instead of a comprehensive approach, cities and counties have initiated policies and programmes to reduce vehicle demand by placing "trip reduction" requirements on employers and office building developers. These programmes are often generated by specific site-based needs and are often implemented as a result of a state requirement, as noted previously. Such policies and programmes have tried to limit or control the amount and price of parking as an effective means of reducing vehicle demand.

Measures and application

The Congestion Management System requirement will be applied to all State level of government regardless if they have an existing comprehensive congestion management policy or program. The Congestion Management System requirements will be added to existing transportation planning and project funding requirements that State governments must follow in order receive funds from the Federal government for congestion management measures.

Institutional and programmatic issues

Draft regulations are being prepared based on comments received from the general public. The following issues were some of the issues raised and are being addressed in the draft regulations:

What is the proper role and responsibility of the city and the county in working with the state to develop the Congestion Management System? How can coordination between the levels of government take place in an effective manner?

111

What are appropriate performance indicators to evaluate and monitor the impact of congestion management measures? Is there one indicator of performance for a congestion management measure or are there several to meet different types of city and state needs? Should the Federal government require an indicator of performance when congestion is viewed differently across the country?

How much data needs to be collected by the State, the city or the county to effectively monitor the impact of congestion management measures?

How does the requirement for the Congestion Management System fit into the existing transportation planning requirements?

Can congestion problems and effective relief measures be modeled?

The congestion management system as envisioned will emphasize supply and demand side measures; the way the combinations of these measures will evolve at the state, city or county levels of government will be the result of the congestion management system requirements, and other factors.

It is essential that the availability of funds be tied to the development of a congestion management plan. On most other factors, it is too early to identify conclusions or successes from the congestion management system requirements.

III.4.12. The Swedish National Committee on Urban Traffic (1989-90)

Objectives

The Swedish National Committee on Urban Traffic have set up the following objectives for the three Swedish largest metropolitan cities :

♦ Reduce air pollutants from traffic;
♦ Reduce the interferences caused by traffic noise;
♦ Reduce congestion.

Measures and application

A considerable number of interacting measures are needed :

♦ Substantial upgrading and expansion of public transport services;
♦ Construction of main traffic routes for solving local environmental problems;
♦ Introduction of car tolls and regional environmental fees;
♦ More stringent exhaust requirements for cars and heavy vehicles;
♦ Increase of fringe benefit taxation for private use of company cars;
♦ Restrict the volume of heavy diesel-driven traffic in city centres;
♦ Increase of the demands on noise levels for vehicles;
♦ Increased parking fines;
♦ Reinforcement of regional and urban master planning;
♦ Constitutional changes for road tolls, regional environmental fees, parking fines, etc.

Effects

The effects of these proposals have been estimated to lead to a reduction of 30-40 per cent of vehicle mileage in the inner city and an increase of public transport vehicle speeds by at least 30 per cent. The environmental goals for the year 2000 is said to be a reduction by 50 per cent for nitrogen oxide emissions and a reduction of dangerous substances of 90 per cent in the city centres. The motorists will greatly benefit from reduced travel time. The savings from an economic stand-point would be significantly greater than the increased transport fees incurred by motorists.

III.4.13. Area Licensing Scheme (Singapore)

Objective and principle

Area wide congestion pricing is the core aim.

Measures and application

The programme, implemented in 1975 and still operating today, charges all passenger cars carrying fewer than four occupants for travel into the downtown priced area from 7:30 to 10:15 in the morning. Today, the fee is approximately $2.50 per day. During the first few years, the fee was $1.30. Fees are imposed by daily licenses which can be bought at roadside stands on approach roads and selected post offices. They can be bought in advance. Monthly licenses are available. Licenses are displayed on windshields. Fifty officers at 28 crossing points enforce the programme. Violators are cited by mail. The fine for travelling without the license was set at $23 for first offense and increase sharply for repeat offenders. Collateral supportive measures included a one-third increase in bus service to the downtown, addition of park-and-ride lots along approach roads, and increases in parking rates in the central area.

Costs

The cost and revenue picture is positive. Initial capital costs were estimated to be less than $500,000. The annual revenues from license sales are estimated to be between $5 and $10 million. Annual operating costs have been estimated to be only a small fraction of revenues.

Effects and effectiveness

The pricing programme reduced peak period traffic in 1983 by 23 percent below pre-pricing levels, and by over 40 percent below the traffic level that would have been reached if normal traffic growth had occurred. Auto-use share of work trips to the area decreased from 56 to 46 percent of total after the programme was implemented (by 1983, this share was down to 23 percent). Bus share went up from 33 to 46 percent. On the other hand, traffic on peripheral bypass roads increased considerably. The afternoon peak congestion was not reduced significantly. The pricing program was extended to PM peak in 1989, all exemptions except for buses were eliminated, and the price was dropped to $1.30 per day. It is expected to move from supplementary licenses to AVI (Automatic vehicle identification system) in the near future.

Two other positive results relate to pollution and business activity. Average daily air pollution fell 10 percent, and pollution in the morning peak declined by up to 30 percent after implementation. The programme apparently did not adversely affect business activity or rents within the area.

Observers cite several reasons for successful implementation of congestion pricing in Singapore. A strong, authoritative government which was generally supported and trusted reduced the chances of protest, suites and opposition. Severe congestion supported the case for strong action. Timing the programme to coincide with significant transit expansion and economic and land use reforms also eased acceptance. Also important was a good public information process stressing transit expansion, pricing of the peak period only, and benefits of reduced congestion.

III.5. REFERENCES

1.   ALLAN, M.G. (1993). *Traffic Initiatives for the Capital. Priority (Red) Routes and the Traffic Director for London.* OECD Expert Workshop on Congestion Management, Barcelona, March 29-31, 1993.

2.   AQUARONE, J.C. (1993). *Objectifs du transport combiné en Suisse.* OECD Expert Workshop on Congestion Management, Barcelona, March 29-31, 1993.

3.   BERMAN, Wayne (1993). *National incident management coalition in the U.S.* OECD Expert Workshop on Congestion Management, Barcelona, March 29-31, 1993.

4.   CAUBET, C. (1993). *Institutional Issues in Future Congestion Management.* OECD Expert Workshop on Congestion Management, Barcelona, March 29-31, 1993.

5.   CETUR (1984). *Les Plans de Déplacements Urbains.* Recueil de texte. Bagneux, France.

6.   DE BOER, R. (1993). *Impact of a new railway line (the Flevo line) on travel patterns.* OECD Expert Workshop on Congestion Management, Barcelona, March 29-31, 1993.

7.   DE BRUIJN, T.E. (1993). *Transportation Region of Groningen.* OECD Expert Workshop on Congestion Management, Barcelona, March 29-31, 1993.

8.   DE MOL, F.J.M., and J.C. VAN DER ZWART (1993). *Congestion Management in the Netherlands.* ITE 63rd annual meeting compendium of technical papers. The Hague, September 19-22 1993.

9.   DIRECTION GENERALE DE LA SECURITE ET DE LA CIRCULATION ROUTIERES (1991). *Exploiter la route, Cadre de réflexion.*

10. DITTMAR, Hank (1993). *Implementing Congestion Management Systems: Confronting and Resolving Institutional Barriers.* OECD Expert Workshop on Congestion Management, Barcelona, March 29-31, 1993.

11. DURAND-RAUCHER, H. (1993). *The Effect of Certified Traffic Information on Driver Behaviour.* OECD Expert Workshop on Congestion Management, Barcelona, March 29-31, 1993.

12. HOPPE, Kurt (1993). *The Importance of Public Transport in a Strongly Ecological Oriented Traffic Policy: The Case of Berne.* ITE 63rd annual meeting compendium of technical papers, the Hague, September 19-22, 1993.

13. INRA (EUROPE) (1991). *Eurobarometer 35.1: European Attitudes Towards Urban Traffic Problems and Public Transport.* Survey Report for The Commission of the European Communities and the International Union of Public Transport (UITP).

14. ITE 63rd annual meeting compendium of technical papers, the Hague, September 19-22, 1993. *Session 29: Road and Trip Pricing Approaches.* Contributions by T.D. Hau; H.D.P. Pol; M.G. Richards; G. Dobias, F. Leurent, F. Papon; T. Bjørgan; D.J. Turner, P. Olszewski; J. Behrendt..

15. KTE. 24th International Scientific Conference in Budapest on Transport Planning and Traffic Engineering about Traffic Management. Budapest, April 27-29, 1993.

16. MAY A.D., and M. ROBERTS (1991). *Demand Management for Urban Road Traffic in the United Kingdom, towards a sustainable transport policy for urban areas.* IATSS Research Vol.15 No1.

17. MINISTRY OF TRANSPORT AND MINISTRY OF HOUSING. *Physical Planning and the Environment (the Netherlands): Second Transport Structure Plan:*
    - *Part A: Policy Intention (November 1988)*
    - *Part B: Summary of Comments from the Public (August 1989)*
    - *Part C: Recommendations from Advisory Bodies to the Government*
    - *Part D: Government Decision (June 1990)*
    Published by SDU Uitgeverij, The Hague (In Dutch, Part A and D also available in English).

18. MOGRIDGE, Martin J.H. (1980). *Travel in towns: Jam yesterday, jam today, and jam tomorrow?* Macmillan, London.

19. OTT, Ruedi (1993). *Traffic in Zürich.* International Conference "Travel in the city - Making it sustainable" .Düsseldorf, 1993.

20. ROTHENGATTER, Werner (1993). *Obstacles à l'utilisation des instruments économiques dans la politique des transports.* Séminaire conjoint OCDE/CEMT sur l'internalisation des coûts externes des transports. Paris, 30 septembre - 1 octobre 1993.

21. SLAGER, Jan (1993). *Presentation of the Corridor Study Amsterdam-Utrecht.* OECD Expert Workshop on Congestion Management, Barcelona, March 29-31, 1993.

22. VAN DER HOORN, A.I.J.M. (1993). *The Dutch Transport Structure Plan 1986-2010, a progress note after three years.* OECD Expert Workshop on Congestion Management, Barcelona, March 29-31, 1993.

23. VAN DER HOORN, A.I.J.M. (1993). *The Dutch Transport Structure Plan 1986-2010.* ITE 63rd annual meeting compendium of technical papers, the Hague, September 19-22, 1993.

24. WOOTTON, John (1992). *Pour une meilleure utilisation de l'espace routier. 17ème semaine internationale d'études sur les techniques de la circulation.* Varsovie.

CHAPTER IV

THE FUTURE OF CONGESTION MANAGEMENT

The previous chapters reviewed current measures and programmes to relieve road traffic congestion. This chapter presents prospects for the (possible) future of congestion management based on perceived trends, constraints and policy goals affecting traffic and transport in the near future with some examples to illustrate future policy elements and considerations.

The potential of new technologies must be assessed against the background of current research and development activities. These have been recently discussed in an OECD report (1) and have led, in 1994, to trilateral co-operation and exchange between Europe, North America and the Pacific. However, the more general challenge and perspective in regard to congestion management are that it is a factor and part of sustainable development of our society with increasingly complex energy and environmental impacts.

IV.1. TRAFFIC AND TRANSPORT IN THE NEAR FUTURE

IV.1.1. Future trends and constraints

In simple terms, congestion is related to traffic volumes that result from cars driving at the same time to the same destinations. The severity of the road traffic congestion problems in the future will therefore largely depend on the growth of *car traffic*, which will be influenced by:

♦ the trend in the demand in road transport; and, on the other hand,
♦ the growing awareness of limitations to growth of car use.

If one considers some of the basic factors which are responsible for the demand in (road based) transport it becomes obvious that there is a sustained trend for growing demand in the near future, i.e. in the next 10-20 years. The countries of the European Union face an average annual growth rate in car ownership of 4 per cent (2) and will in a few years reach the present level of the United States which is about 0.6 private cars per capita (3). Similar trends have been identified in the other OECD countries. With the growing availability of the private car, the overall demand for driving car will increase.

There are demographic changes which will influence the future demand in road transport. For example in Europe during the next two decades the number of people older than 64 years will increase by 20 per cent (4) and in the United States, the per cent of individuals 65 years of age or older is expected to double between 1990 and the year 2030. In contrast to the present situation, it will be a matter of course for elderly people to use their private car as long as possible.

Since 1965, the number of U.S. women with driver licenses has doubled; the availability of a license is about 85 per cent for women, contrasted to 77 per cent in 1983. For every 1 per cent shift from non-drivers to drivers in the female population, total travel in the U.S. jumps almost 10 billion miles per year.

Experience in the United States indicates that the number and per cent of workers in suburban areas is increasing faster than in the central cities of metropolitan areas of more than one million population. In many major metropolitan areas over 50 per cent of workers live and work in suburbs.

Up to now, no changes in lifestyle with respect to the use of the private car are visible. For many people it seems to be the most convenient way to travel by car from A to B both for business and more and more for non-business purposes.

These trends support the hypothesis that as long as there are neither enormous improvements in alternative means of transport (such as public transport or ridesharing), nor strong restraint for private car use, the growing demand in private car usage will continue and congestion problems will spread. (However, in the U.S. trends show that even with improvements to public transport systems people will not give up there private cars without further incentives).

A similar trend of a growing demand can be expected in *freight transport* which, for example in the European Union, has increased during the last 20 years by about 50 per cent (4). Most of this amount has gone to road transport, motor trucks, because it is in many cases faster, cheaper and more flexible than the competing systems. The current European modal split between the road and other modes of freight transport, based on tonne-kilometres, is about 70 to 30 (4). The current U.S. modal split, based on tonne-miles, is about 26 to 74 but truck transport is on the rise. New principles of manufacturing like just-in-time production with the demand for punctual delivery require efficient transport facilities and, therefore, intensify the demand for road based freight transport in the future.

Freight transport is the backbone of our free-market economy. Therefore, the total demand of freight transport will not decrease; a further increase can be expected, e.g. by new trade relations due to changes in the political and economic systems in Eastern Europe or the North American Free Trade Agreement between the United States of America, Canada, and Mexico. It has been estimated that a 1 per cent increase in Gross National Product (GNP) is associated with a growth of 1.5 per cent in passenger transport and 3 per cent in freight transport (2). In the United States, passenger and freight transport expenditures account for about 15 per cent of the GNP (5).

Contrasting to the trend for growing demand in road transport there is another trend visible: the limitations to growth of *car use*. Car use is known to have not just positive effects on the mobility of the individual, but also to have negative effects at a variety of societal levels and scales.

At the smallest scale of the 'neighbourhood' (residential areas, city centres with a number of functions, areas with mixed functions) the growth in car use and parking will meet certain limitations, since liveability and accessibility will be largely affected. Exhaust fumes, noise and the vehicles themselves influence the attractiveness of the living area and the health of the inhabitants. Particularly the scarcity of space will restrict further growth of car use in neighbourhoods. At the level of the municipality (towns, cities), the growth of car use will be restrained because of limited space -- which can be more efficiently used for other functions -- requirements concerning the attractiveness of a city (for shopping, visitors, settlement of new industries, etc.) and costs for facilitating car traffic (investment for and maintenance of roads). At the national level, increasing car use will be in conflict with ecological standards -- since new road connections rapidly decrease the living conditions of numerous species in nature -- as well as acceptable norms of nitrogen oxide (NO_x) pollution that soil and water can bear with respect to acidification and the number of road accident victims society is prepared to accept. Finally, at a global scale the limits of emissions of carbondioxide (CO_2) and consumption of energy could largely determine the acceptable levels of car use in the future.

Furthermore, public awareness of the limitations of a careless exploitation of 'Mother earth' is increasing. One of the main issues is environmental awareness, but also the conservation of nature and landscape are of major concern. Public awareness and concern about air and noise pollution and about global warming have led to the enactment of major laws and policies that will have a strong influence on the way in which transport decisions are made. For example, the Clean Air Act Amendments of 1990 affect transport investment decisions in the United States; they also require the private sector to take on much more responsibility for providing transport service, especially for their employees. Environmental issues (noise, air pollution, global warming, etc.), safety, land use, congestion, etc. will certainly have a strong influence on the policies and measures applied to address traffic problems in the future.

Another feature in a number of cities in OECD countries is the growing consciousness of citizens and politicians that most of the cities cannot bear further growth of car use. Can the sound functioning of a city be reconciled with a growing role of the car in the city? Solutions like introducing areas where car use is restricted, shifting parking locations to areas further away from the city centres together with improvements of transport by transit and bicycle, are becoming common in many cities. Both environmental and congestion issues are addressed through such approaches; they mark the awareness of both citizens and local politicians that alternatives for the abundance of the car in the city are not just possible, but can be functional and attractive as well.

IV.1.2. Future transport policy goals

It is clear from the evidence gathered by the Expert Group that future activities in congestion management must be supportive of and compatible with broader *transport policy goals* that include:

- ♦ a guarantee of reasonable levels of service, mobility, and accessibility.
- ♦ support for economic growth and productivity (freight and passenger movement).
- ♦ support for environmentally sustainable development.
- ♦ improvement of traffic safety.
- ♦ balance between demand and supply for different transport modes, services, and facilities.

Achieving these goals against a background of space limitations, environmental concerns, and financial constraints will be the real challenge for transport professionals in the future. It can no longer be business as usual - that is providing more road infrastructure for a given transport demand.

Based on the work done by the expert group, the principal *congestion management policy objectives* to reduce traffic problems -- and to meet, at the same time, the much broader transport goals -- are likely to be as follows:

♦ **Maintain adequate mobility through provision of acceptable alternatives to the private car** (public transport, car pooling, bicycle/walking, etc.),

♦ **Alter trip patterns** (land use, alternative work schedules and/or telecommuting, travel behaviour etc.),

♦ **Improve traffic flow** (route guidance, traffic control improvements, incident management etc.)

These policy goals are in line with the assignment of impacts of congestion management measures presented in Tables II.3 and II.4.

There will be different policies in the OECD countries to reach these overall goals depending on the different geographical, political, social, cultural and financial situations. Some policies are common for many countries, others are restricted for a few. The following Section IV.2 gives examples for future national policies on a general level based on the national contributions to this report.

IV.2. EXAMPLES FOR FUTURE POLICIES AFFECTING CONGESTION MANAGEMENT

In a pluralistic society, a long time is needed to practically implement policies in terms of a set of measures. The different "actors" -- government, administration, industry, users, inhabitants, etc. -- have their own, often conflicting interests and are therefore involved in the process of formulating policies. Therefore, it is difficult to make accurate forecasts as to the implementation of future policies in the transport sector. The following examples illustrate the trends in policies to reduce the traffic related problems of the future in different countries. They are quoted to show that congestion control and demand management measures are important tools within an overall transport policy.

IV.2.1. Expanding infrastructure

It was already mentioned in Section IV.1 that there will be more and more problems to extend the existing road infrastructure in OECD countries. In the United Kingdom, the long period of planning, land acquisition, financing and construction imparts considerable long-term momentum to the national road construction programme. However, most of the new work is concentrated on up-grading existing heavily loaded roads, or on building motorways to relieve traffic on nearby all-purpose roads suffering from severe overloading. Growing public resistance, often leading to appeals to the European Community Institutions alleging inadequate prior provision of Environmental Impact Statements, makes it increasingly difficult to secure public acceptance for opening up entirely new routes.

In the western part of Germany road construction is for similar reasons restricted to extension of existing roads (e.g. 4 to 6 lanes) and to short new constructions for the completion of the network. The construction of new roads cannot be the future mission of congestion management due to limited space in the densely populated country, environmental issues and a public opinion generally against road building.

A similar situation exists in the Netherlands where a better exploitation of road capacity must be realised by tailoring to specific situations: possible expansion of the road infrastructure will need to be considered for each transport corridor on its own merits, case by case and in conjunction with possible development of railway and tram networks. The accessibility of the intercontinental ports will be a central focus of the Dutch transport policy.

France is an example of a country which -- although facing the same problems in urban areas -- reserves enough space for building new infrastructure in several regions outside urban areas. However, against the background of the limitation of possible investments (decreasing budget of authorities, complexity and risks for private investments) and the phenomenon of newly induced trips, it will be more difficult to build new roads.

In the United States, the expansion of existing roadway infrastructure will be done within the bounds of Federal and State environmental goals and policies. Most expansion of the existing infrastructure will be done primarily for the purposes of safety improvements or to accommodate high occupancy vehicles (HOV's) such as buses, carpools, or vanpools. The emphasis on expansion of the existing infrastructure has been replaced by emphasis on better management of the existing system through operations as well as economic and technological improvements (known as congestion management).

The growth of new roads in the United States will be limited because of high costs, constrained resources, availability of land, environmental concerns, and developmental issues. Efforts are underway in States like California, Colorado, and Virginia to enable the private sector to build and own new roadways that are offered to the public for a price. In addition, the new roads that are constructed often have tolls placed on them to help pay for the cost of construction and maintenance.

Another important issue which limits the extension of the road infrastructure is the necessity to use the future road infrastructure budget for the most part for maintenance and renewal of existing infrastructure. Especially expensive constructions like 20-30 year old concrete bridges have to be restored in the next decades.

IV.2.2. Environment

Another important policy is the protection of the environment. New technologies to restrict the emission of injurious gases (using substitute fuels or new solutions for vehicle motorisation) will be developed in response to progressive tightening of laws and regulations, but this takes time.

The consideration of environmental issues gains increasing influence in all discussions concerning traffic. It seems most likely that no measure for congestion management will be considered in the future which has more negative effects on the environment than the current situation. However, positive effects of measures on the environment (and traffic safety) could constitute an important argument for implementing measures (possibly more important than arguments like improved efficiency or congestion management!).

Up to now there seems to be a different attitude towards the importance of environmental issues. In Germany there is a strong political intention to reduce the carbon dioxide emissions (responsible for the greenhouse effect) by 25 per cent by the year 2005. Within the transport sector the reduction should be achieved by a package of measures like reducing the travel demand (pricing policies) or the enforcement of 'ecologically harmless' transport modes. For example, in France the environmental issue has -- until now -- not reached similar priority levels such as to say that traffic reduction is needed, but certainly this will be the case in the future.

The United States, like other OECD countries, will address environmental issues in the future development of its transport programmes. A unique feature of U.S. policies in transport and environment is that funding of transport projects (like congestion management) is tied to compliance with environmental laws. For example, national funding of transportation projects under the Intermodal Surface Transportation Assistance Act of 1991 is allowable only if States and urban areas meet the requirements of the Clean Air Act Amendments of 1990. Most States have similar relationships between transportation funding policies and environmental laws.

IV.2.3. Economic policies

At the moment another policy for reducing travel demand is under discussion: the enforcement of economic pressure (today, road pricing is mostly not used as a measure for restricting travel demand but for raising funds, which can be used for other investments in transport infrastructure). There is the long term intention to charge real costs of transport especially by modulated road tolls and -- perhaps -- by increasing public transport fares. The latter shows, that pricing policies must be tackled in a careful way because of social reasons. Transport facilities must be accessible to all people.

Presently, road pricing (in some cases also referred to as congestion pricing) has been introduced in a few countries but at a limited, localized scale. A reason for the limited projects in Europe is the European rule of "equal treatment". This means that all countries in the European Community must be treated identically. Therefore, road pricing programmes are not allowed, which do not take into account the distance or time travelled (e.g. like the vignette in Switzerland, which must be bought for the whole year). Toll stations, like in France or Italy, cannot be used for example in Germany where the highway network is dense with many access points. This brings about the conclusion that the future of road pricing lies in automatic electronic debiting systems. If these systems are available, and market proved on an international standard, road pricing could be an important means of transport policy in the near future - provided that no long delays will occur in the gradual adjustment of policies and charges within Europe. In case of parking pricing, the positive effects on urban car traffic are already visible. Therefore, parking pricing will remain an important tool for managing urban car traffic in the future.

IV.2.4. Freight transport

To reduce freight transport on roads or to promote the freight share of alternative modes different policies are considered in Member Countries. Switzerland, the Netherlands and Germany for example invest substantially to improve their railway system in order to accelerate (passenger and) goods transport. However, it is uncertain whether the transport time savings obtained will be sufficient enough to influence the modal split in goods transport.

Due to the fact that the share of road transport costs for goods is only 5 to 10 per cent, it is unlikely that road pricing will reduce road freight transport in the near future.

In all countries with heavy truck through-trips like the Netherlands, Germany, Switzerland, France, there is a strong declared intention to improve combined transport (also called intermodal transport) facilities not least because of environmental and economic reasons; however, in most cases and with the exception of Switzerland, there is no real consensus, nor a political programme. The key to increase the existing combined transport facilities lies largely with logistics, where new developments in telematics and in the definition of electronic data interchange procedures may in time make a substantial contribution. It will however be necessary for railway authorities to give this subject a greater priority than present tentative development studies currently imply.

From a congestion management perspective, the future freight movement policies in the United States will be characterised by intermodal features. Major road corridor studies along the East, Midwest, and West include provisions for improved truck flows and for better access to ports of shipping (rail, water, and air terminals). New policies presented under the Intermodal Surface Transportation Efficiency Act of 1991 will mandate that cities and States incorporate freight movement activities into their congestion management and overall transport programmes. Freight movement measures will be incorporated into congestion management activities as a matter of national and local policies.

IV.2.5. <u>Land use</u>

Land use policies must be developed and encouraged in order to reduce car use and trip length. Offering decent alternatives will not get enough people out of their cars unless negative stimuli are also applied. It must be ensured that the number of kilometres travelled in both goods and passenger transport is reduced, for example by reversing the trend towards ever-greater travelling distances between home and work. This will involve not only measures to influence people's behaviour but also a carefully balanced location policy for housing, work and recreation.

Current land use policies that facilitate uncontrolled suburbanisation will be revised in light of growing environmental, traffic, and growth concerns. Future trends in land use policies that manage or reduce congestion include pedestrian/transit friendly environments, mixed use developments, parking restrictions, growth management, etc. However, reversing the established patterns of separate housing and business areas will be a complex and long process.

IV.3. NEW TECHNOLOGIES

The policies presented in Section IV.2 require political, administrative, financial and technical efforts and support so as to carry them out successfully. Chapter II presented many different measures to carry through the various individual policies. These measures are likely to be used in the near future, hopefully in a more comprehensive approach than stated in Chapter III. Moreover, each single measure will be improved by further technological advancements. For example, new technologies hold the potential for improved traffic control, guidance and surveillance, faster communications between the driver and the transport agency, a more informed traveller due to the use of (real time) accurate, localized data and better enforcement of traffic laws.

Advanced technologies in the fields of informatics, telecommunication and vehicle manufacturing provide the basis for a better road transport system in the future. Therefore, the role of new technologies will briefly be outlined in this section.

Most of the countries represented in the study group invest in research on advanced technologies to improve congestion management. Looking at the current R&D programmes concerning advanced technologies, like IVHS-America in the United States, PROMETHEUS and DRIVE in Europe or VICS (now ARTS and VTMS) in Japan, one could get the impression that the future of congestion control

and demand management has just begun. All programmes started in the end of the 1980s with the high level objectives to:

- improve road safety,
- maximise (road) transport efficiency and
- reduce environmental impacts,

using the possibilities offered by innovative technologies in the fields of informatics, electronics, telecommunication and broadcasting. The programmes reflect the competing situations of the United

States, Europe, Australia and Japan to develop new technologies called IVHS (Intelligent Vehicle Highway Systems), RTI (Road Transport Informatics) or ATT (Advanced Transport Telematics).

The following examples of ATT-applications and services -- classified by area of major operational interest, as established in the DRIVE II programme -- should give an insight into the current activities in the field of IVHS/ATT which is the basis for future technological advancement in congestion management:

- Demand Management: Transport demand should be influenced by technologies and strategies for controlling the use of road space, access control and parking management/pricing (e.g. automatic debiting systems, multi-modal payment systems (smart cards), etc.).

- Travel and Traffic Information Systems: This area covers all activities for collecting, processing, predicting and distributing (dynamic) travel and traffic information. The user (car driver, public transport user/operator, fleet manager) receives the information at home, in office, on street or into the vehicle allowing him to make better choice about travel time, mode and route.

- Integrated Urban and Interurban Traffic Management Systems: Work in these fields should improve the performance and integration of transport systems by new techniques in on-line data collection (e.g. O/D flow estimations, automatic incident detection, traffic, weather and road condition monitoring) for advanced traffic management and control (like collective and individual route guidance, advanced signal control, travel and traffic information, parking management, emergency management, incident management, environmental control, weigh-in-motion systems etc.).

- Driver Assistance Systems: These systems – which will be available further in the future – represent innovations within and outside the motor vehicle to assist the driver for example with collision avoidance systems, on-board road signing and visual enhancement, "car-train systems", vehicle-vehicle communication systems, driver information systems, etc.

- Freight and Fleet Management: In this area the use of advanced information and telecommunication techniques for improving freight operations with respect to logistics, fleet and vehicle management are developed and tested. Major emphasis is given to the use of mobile communication (radio, satellite, cellular radio, microwaves) and access to information services and databases.

- Public Transport Management: This area covers all activities in the field of public transport like vehicle scheduling and control systems, dynamic user information systems (at home, at stop, in vehicle), demand responsive services, advanced payment systems, integration of public transport into urban traffic control, etc.

Each R&D-programme builds on a framework of concertation to develop a common approach for the different groups involved like users, industry, research institutes, service providers, operators and authorities. Although the programmes differ with respect to responsibility, funding, scope and organisation, they have similar tasks which can be described as follows:

- Development of ATT-applications for all areas described. New system components are developed as prototypes and/or are improved up to marketable products.

Traffic Control Center
(Barcelona, Spain)

Traffic Control Center
(Tokyo Metropolitan Police Dept.)

♦ Research and development activities in supporting areas (like transport modelling, behavioural aspects, databases, man-machine interfaces...). Results are used as building blocks for further ATT-development and implementation.

♦ Operational field tests of ATT in real world situations in order to evaluate applications of new technologies and system concepts. Field tests are used to gather data on real-world costs, benefits, operational performance and reliability. Furthermore, field test help to assess institutional issues, market support and user reactions. Field tests should also promote confidence in ATT-services amongst all parties involved.

♦ Significant contribution to standardization and common functional specifications especially with respect to interface requirements. Commonly defined interfaces allow for interchangeability among devices and systems from various suppliers. Standards reduce the risk for manufacturers when developing ATT-products. There is a particular need for European standardisation in order to facilitate the use of ATT-services across all countries. Consequently, the provision of directives and guidelines to which the ATT-products and the intelligent transport infrastructure should conform are further outputs of the R&D programmes. From the point of view of maximising the efficiency of production and distribution of equipment, there is a strong case also for international standardisation, and the various national and sectoral interests are addressing these activities through new Technical Committees at both European (CEN/CENELEC/ETSI) and international (ISO,IEC,ITU) levels.

The conceptual and operational frameworks provided by the R&D programmes in the United States, Europe and Japan concentrate the national or regional efforts in the field of advanced transport telematics and, therefore, minimise unnecessary duplication of efforts by the help of various concertation processes.

The fault the critics find with ATT-programmes is that the emphasis is put on the (private) automobile. They argue that car driving increases with the provision of better road facilities. Indeed, just at the beginning of the programmes, public transport concerns were not sufficiently represented when compared with their important potential in reducing road traffic and congestion.

However, this unbalance has somewhat changed during the course of the realisation of these programmes. Their structures have changed since the starting phase of exploring the options of ATT up to the conduct of operational field tests which require an integrated approach, not least due to institutional reasons. In carrying out these field tests, the requirements specified by the infrastructure providers, i.e. local authorities and politicians which are responsible for the implementation and financing of services and systems, are very important. Today, in a society of increasing awareness of environmental issues, services which only favour the private car cannot be installed without political resistance. Therefore, public transport and park & ride facilities gain gradually more importance in the real world environment. This is one reason for the inclusion of advanced public transport services in many operational (urban) field tests.

At the moment, the ATT-programmes are roughly in the middle of their schedule (e.g. IVHS: 1997, DRIVE II: 1994), extensions are expected. Operational field tests have just begun. First results are due at the end of 1994. The programmes in Europe are generally at the stage of technological validation rather than full-scale trials that are appropriate for measuring total social, behavioural and economic impacts. This is a role to which the Community's Fourth R&D Framework Programme could make a major contribution.

With respect to demand management and congestion control it can be expected that the new technologies developed within the ATT-Programmes will improve most of the individual measures described within Chapter II in the near future. Due to the integrated approach of many field tests, findings with respect to comprehensive measures will be available. However, no miracles can be expected by the implementation of new technologies. They can only be technical tools in a comprehensive approach for which the requirements are already described in Chapter III. A comprehensive approach involving authorities, operators, service providers, industry and users seems to be the only way for reducing the traffic problems of the future.

IV.4. SUSTAINABLE DEVELOPMENT POLICY

Congestion is not an isolated problem. Congestion is a feature of the 'modern Western' society, which is largely car dependant. In a sense, car dependency leads to an increase of car use. In a sense, car use is part of the more or less destructive life style of this society.

World leaders agreed in Rio de Janeiro in 1992 that the continuation of this destructive life style of the west, and a threat for copying this life style by a number of other countries in the world, eventually goes far beyond the limits the earth can bear. These limits, although they are not yet quantitative, are related to the notion of 'sustainable society'. A sustainable society stands for developments which satisfy the needs of the present society without endangering the possibilities for future generations to satisfy their needs as well. World leaders are aware that limits need to be set to a number of negative aspects of modern living, for instance the use of fossil fuels and other raw materials, the emission of hazardous gases like carbondioxide and CFK's, the spilling of nature etc.

A general strategy for 'sustainable development' covering all sectors of our society should assure that energy and material flow should be reduced and the used resources be renewable. Production process must be as clean as possible and the end-made products should be actually recyclable. This strategy should be forced by Push-and-Pull measures as a combination of:

1. Laws and regulations in order to protect or hinder misuse of resources and promote use of renewable energy and recirculation of material.

2. Pricing of the use of products and the infrastructure in order both to promote a desired behaviour among consumers and to finance investments for environmental protection.

3. Publicity and information in order to influence and promote understanding of the desired strategy.

4. Co-ordination of the public and private sectors in order to establish and implement actions.

The transport sector will be of certain interest for 'sustainable development' in the future as it consumes a great deal of non-renewable resources in energy and raw-materials and as the negative environmental impact has pervasive repercussions both on the local and the global level. The transport sector in industrialized countries stands accounts for 15-20 per cent of the total energy consumption, of which 2/3 is attributable to the use of private cars. There is today no immediate sign that the demand for road traffic will decrease. This means that the negative impact of transport on the environment will not only persist but may worsen.

In view of increasing congestion, unsafety and environmental problems associated with limited economic resources, the current transport policies and plans for increasing mobility by cars and trucks must be questioned, even if there are ongoing programmes for less oil consumption, reduction of noise and exhaust gases, material recycling, etc. As there may be no time to lose, the main issues must be: How to reduce car dependency, how to change the behaviour of motorists, and how to offer alternative modes of transport, destinations etc.

Steps towards sustainable development coincide with the goals of reduction of road traffic congestion as described in this report, especially on the demand side. However, the methods should --- to be in line with 'sustainable development' -- be more directed to reduce private car dependency and car traffic and may be stronger in application, e.g. through regulation and pricing principles. It is therefore possible that a future transport policy towards sustainable development could cover the following items:

♦ Any increase of existing road traffic volume should not be accepted.

♦ Plans and projects, aiming at increasing road capacity, should be stopped. Plans for land use change should be directed to reduce car traffic and environmental effects. If they are increasing car traffic or causing environmental impact, they should be stopped.

♦ Alternatives to car mobility should be offered. Public transport service, based on pro-environmental vehicles, should be increased and promoted. Also, effective use of vehicles (car and van pools) should be supported.

♦ Road tolls should be introduced both to regulate demand and to finance investment in alternative modes of transport.

♦ Publicity and other information measures should be directed to individuals and groups concerned in order to explain positive and negative effects of the measures.

♦ Co-operation with the public and private sectors will be necessary for creating acceptance and support for implementation of measures affecting e.g. travel of employees to and from work, freight and delivery transport.

Sustainable developments gain increasing importance within the transport sector in urban and traffic planning (e.g. "Transport Structure Plan" for The Netherlands). However, there is no programme strongly and actively working towards sustainable mobility in a more wider sense. The problem is of course the difficulties to convince people that the freedom of mobility must be revised. Even if e.g. the congestion problem is evident and visible, the global environmental problem is not.

In the future, research and demonstration projects are necessary to find out ways to convert today's transport behaviour towards sustainable mobility. Also, problems of the decision process must be taken up as there will be a lot of conflicts between need for mobility, accessibility and reducing the negative impacts of transport. It is also necessary to get more studies on the behaviour of transport consumers (individuals and companies), e.g. how transport demand is created under various conditions, how transport strategies are developed and carried out, consumers' reaction to various measures to influence demand etc. The studies should be related to the role of all actors involved in the transport sector, such as politicians, authorities, transport industries, transport operators, real estate and construction companies, commercial and business companies, lobby groups and all others.

As long as the political will is still too weak for unpopular decisions and many transportation professionals believe that the transport and traffic problems can be solved exclusively by technical means, it may be questioned if these problems can be solved within the transport policy sector at all.

IV.5. REFERENCES

1. OECD (1992). *Intelligent vehicle highway systems: review of field trials*. OECD. Paris.

2. COMMISSION OF THE EUROPEAN COMMUNITIES (DGXIII) (1993). *Transport Telematics 1993. Brussels*.

3. U.S. DEPARTMENT OF ENERGY (1992). DATA BOOK, Edition 12. Washington D.C.

4. COMMISSION OF THE EUROPEAN COMMUNITIES (1992). *Transport Telematics Requirement Board; Telematics applied to transport (DR 921780)*. Brussels.

5. U.S. DEPARTMENT OF TRANSPORTATION (1992). *National Transportation Statistics - Annual Report, June 1992*. Washington DC.

CHAPTER V

CONCLUSIONS AND RECOMMENDATIONS

V.1. SUMMARY

This Report was prepared by the OECD Expert Group on "*Congestion Control and Demand Management*" to share information between Member countries on a broad collection of measures applied to reduce the impact of road traffic congestion problems. The Report is designed to show what conventional and innovative measures are being taken to address the increasingly difficult traffic congestion issues.

The collection of measures are termed "Congestion Management" and presented in two strategy classifications: demand-side and supply-side. Demand-side congestion management measures are designed to reduce car demand on the system by increasing vehicle occupancy, increasing public transport mode share, reducing the need to travel during a specified peak time period, and/or reducing the need to travel to a specified location. Supply-side congestion management measures are intended to increase the existing capacity of the system in order to improve traffic flow for all modes. Congestion management measures and strategies offer the opportunity to improve road traffic conditions and reduce demand for car use.

The Report presents a catalogue of congestion management measures in Chapter II. Nearly forty conventional and innovative congestion management measures are presented in nine sections representing nine strategy classes. Within each section, the strategy class along with the associated measures are described, its objectives and major impacts stated, the application and institutional responsibilities are discussed, and any special or associated problems presented. Examples of the application of the measures are presented at the end of each section.

Chapter III presents an overview of the policies, plans, and programmes that enable congestion management measures to be effectively implemented. Case studies are presented as evidence of current experience. Chapter IV refers to a range of socio-economic policies -- environment, economy, and advanced technologies -- to give a picture of how future congestion management measures may need to be developed and shaped in order to be effective.

Group Members prepared technical information from their countries to meet the desired requirements for Chapter II, III and IV. The information was compiled and organised by the respective chapter rapporteurs. Information used to prepared Chapter III was also supplied from the experience presented at the OECD Expert Workshop on Congestion Management, held in March 1993 in Barcelona, Spain.

V.2. CONCLUSIONS AND RECOMMENDATIONS

Based on the extensive amount of information compiled and reviewed for drafting this report, the Study Group agreed on the following conclusions and recommendations regarding the application of congestion management measures in the OECD countries.

1. Road traffic congestion can be better managed

Taking due account of changing demographics, land-use patterns, economic and trade developments and increases in vehicle ownership, the Study Group believes that road traffic congestion problems can be better managed, but not necessarily solved. While there are experiences where congestion problems have been solved at a particular location, the likelihood of finding a specific measure that will give one long-term solution is small. Extensive experimentation with congestion management measures indicates that they can effectively provide congestion relief for only a period of time. The measures have been found to effectively achieve any or all of the following desirable congestion management objectives:

- ♦ to reduce the need to make a trip,
- ♦ to reduce the length of a trip,
- ♦ to promote/encourage the use of non-motorised transport,
- ♦ to promote/encourage the use of public transport,
- ♦ to promote/encourage the use of carpooling,
- ♦ to shift peak-hour travel,
- ♦ to shift travel from congested locations, and
- ♦ to reduce traffic delays.

How effectively these objectives are achieved depends not only on the types of measures implemented, but also on the level of commitment given to congestion management by the public and/or the private sectors. Effectively meeting the objectives means that the operators of road and public transport systems must always monitor areas of congestion even after the measures have been implemented. If necessary, existing congestion management measures may need to be modified or new ones implemented in order to continually achieve the desired objectives. Effectively addressing a road traffic concern may require a continuous commitment of resources beyond that of just implementing a measure and assuming that the problem is solved.

There exists an extensive number of measures that can be applied (by both the public and private sectors) to address the problems created by road traffic congestion. The study group concluded that there is not one "recommended" approach to manage the road traffic congestion problems that countries face. There are many options and varieties of measures that can be taken to address the problems. When applied systematically or in coordination with other measures and with strong governmental support, these measures have the potential of being effective, regardless of their level of sophistication or technological development.

2. Low-cost, conventional measures can be effective

The study group also found that conventional measures can make a significant impact at relieving a traffic congestion problem. Some of these measures can also be low-cost.

While many countries have strong interests in promoting advanced technology applications for congestion management measures, it is the conventional, low cost, techniques that represent some of the most effective measures used to address the problems of road traffic congestion. For example, enforcement of parking restrictions, retiming or improving traffic signal operations, improvements to intersections, enhancements to express bus services, and improvements to pedestrian and bicycle facilities were among the conventional, low cost, measures often used for congestion management. The potential impact and effectiveness of these types of measures must be considered in the development of any road traffic congestion relief programme.

3. Pricing techniques can be effective in congestion relief

The use of pricing techniques to affect travel behaviour and car use was found to be gaining momentum by governmental agencies as an access restriction or congestion management measure in many countries. Congestion pricing programmes can be applied to modify car demand at certain time periods and to generate additional revenues. Up to now, only limited examples are in place; by using pricing techniques the schemes address the problems of road traffic congestion, and raise revenue for construction, maintenance, and operations.

Several OECD countries are currently experimenting with a variety of pricing techniques and methods as congestion management measures. The pricing techniques have been designed to influence a traveller's choice of mode, choice of route, and/or time of travel. They are applied at parking locations, along congested roadways (sometimes called congestion pricing), and in public transport.

Parking pricing techniques have generally been designed to make the cost of driving alone high, in order to encourage more use of carpooling, public transport, walking, and bicycling. In addition to helping achieve these objectives, road pricing techniques are also intended to encourage travel during non-peak time periods and along less congested routes. Roadway pricing techniques are typically administered at toll collection facilities; however, several countries are experimenting with advanced technologies to collect a toll without requiring the vehicle to stop.

Pricing techniques on public transport are designed to reduce the cost of the service to the user. Several countries are establishing weekly, monthly, or yearly commuter pass programmes that have employers and employees share the cost of the transit fare. This type of pricing technique helps to create an incentive for commuters to use public transport and can work in coordination with parking and road pricing programmes.

4. Public support of congestion management measures is essential

Road traffic congestion problems and the application of relief measures confront travellers with conflicts between their own interests and those of the community. Actions to relieve the problems of road traffic congestion (congestion management) have traditionally fallen upon the public sector to implement; however, the success of the government actions depends in large part upon gaining support of the public. Successful congestion management measures have been perceived by the public as being a "fair" way to meet the overall interests of the community.

In many OECD countries successful congestion management measures have been subjected to an extensive public relations process at the planning, development, and implementation stages. It has been found that public relations is an important factor that can lead to success in congestion relief programmes. For example, because of extensive public relations programmes in several countries, preferential treatments (such as bus lanes, carpool lanes, and bus signal pre-emption) are becoming more acceptable to enhance public transport, bicycling, and carpooling operations in congested areas.

5. Traveller information is important to congestion relief

In many countries, traveller information services, either pre-trip or en-route, is becoming an essential part of a congestion management plan. Changeable (or variable) message signs to provide information to the traveller are being implemented on a wider scale in many urban areas or along intercity corridors.

Pre-trip traveller information to provide the latest road conditions and public transport arrivals is also becoming part of congestion management programmes, especially at major employment sites or shopping centres. This type of information service can be provided through telecommunication services at home or in the office, television monitors at kiosks, FAX machines, or voice mail services. Traveller information services are considered to be an important aspect of advanced technology programmes for congestion management.

6. Coordination is an essential aspect of congestion management

Successful and effective congestion management measures are often characterised by good institutional and organisational coordination and cooperation. In many effective programmes, different agencies and organisational units have joined forces and cooperate with each other in order to make the congestion management measure or policy achieve its desired goals. Such co-ordination between agencies and within operating organisations often lead to implementing effective congestion management measures.

7. Congestion management efforts need to start small then grow

Large and complex congestion management measures, such as those used on motorways often started as smaller demonstration projects that were successful, then grew. The study group found that it may be politically more acceptable for a congestion management measure, especially one that requires an innovative policy or technology, to start small and expand into a larger project. Especially for innovative projects that incorporate advanced technology, such a procedure helps the operating agency gain experience, support, and credibility in applying the congestion management technique.

8. The private sector has a role in congestion management

The private sector should be invited to participate in implementing road traffic congestion management programmes, in addition to its traditional role as consultant or contractor. Historically, they have had a strong economic interest in programmes to improve the movement of people and freight. In light of the development of advanced technologies, the private sector will also be playing a role in the development of system operating procedures, architectures, standardisations, and services.

In addition, the private sector is being required in several countries to assume part of the responsibility for reducing car use and road traffic congestion. Much of the motivation for such requirements result from severe congestion problems and environmental concerns, such as air quality or land-use. For example, air quality laws are being established that are requiring employers to develop measures to get their employees to travel to work in public transport and carpools. The private sector is also being involved in the financing of road and traffic signal improvements in order to help reduce the impacts of congestion, especially at their employment site. Much of the involvement of the private sector in congestion management programmes has been a result of some governmental regulation; however, there are significant examples of where the private sector voluntarily became involved in congestion management, because they saw a community or economic benefit.

9. New policies and laws are needed for congestion management

New policies and laws are often needed to implement congestion management measures. This is especially true in countries that are imposing restrictions, controls, or modifications on car use for air quality, land use, congestion, or social purposes. It is also necessary to modify existing policies and laws that could act as barriers to implementing innovative congestion management measures and advanced technologies. Many new policies and laws stress coordination of the agencies and organisations involved in order to make the implementation of the measures successful. Some of the congestion management measures that have been found to be supported by new policies and laws include carpooling, congestion pricing, auto restrictions, land-use and zoning, accessibility to commercial development, and alternative work arrangements.

It is recommended that governmental agencies revise laws, regulations and policies to address these issues in a comprehensive manner. While the technology does exist for congestion relief, it is often the legal, institutional, or financial barriers that have prevented the implementation and support of effective, innovative, congestion management measures. Laws, regulations, and policies need to be reshaped to help build coordination and cooperation and to address contemporary needs and issues more effectively.

10. New technologies will offer tools for congestion management

New technologies, such as for route guidance, traffic management and vehicle identification, as well as telecommunications, hold the potential for helping to ease the problems of road traffic congestion. They offer new tools for congestion management. Based on the progress of activities and developmental schedules, the expert group noted that such technologies are likely to offer benefits for congestion management over the next 5 to 10 years.

11. Accessibility Must be Maintained with Congestion Management

Several countries are developing plans to place restrictions on where and when cars can be driven, particularly in the city centres. Given this contemporary trend, it is important to incorporate the accessibility needs of the cities into these congestion management plans. While car traffic can be restricted, accessibility to major centres of work, shopping, and entertainment must be improved through public transport, walking, and bicycling. Congestion management and accessibility issues need to be addressed as part of a comprehensive programme for meeting the goals of the community.

12. Evaluations are needed in congestion management

It is recommended that a strong emphasis by OECD countries be given to conducting evaluation and demonstration projects in order to improve the knowledge of the impact and effectiveness of the many congestion management measures. The OECD study group noted the lack of quantitative data on the effectiveness of congestion measures. Information gained from evaluations, research, and demonstration projects are essential to the development of effective congestion management measures. Results must be documented in order to share the experiences and good practices of congestion management measures.

13. Training in the practices of congestion management is needed

It is recommended that training courses be established to help transport professionals understand the need for effective congestion management practices. In addition, case studies that will illustrate the principles and concepts of congestion management need to be published and presented at seminars.

V.3. POTENTIAL TOPICS FOR FUTURE OECD EXPERT GROUPS

Based upon the work of the OECD present Road Transport Research Group "*Congestion Control and Demand Management*", additional expert study groups on four specific aspects of congestion management were found to be needed. It is recommended that additional expert study groups investigate and document the state-of-the-practice; current and future issues; and technological measures for the following four topic areas:

1. Road Traffic Incident Management

2. Parking Management including Parking Pricing

3. Road/Congestion Pricing

4. Telecommuting

ANNEX A

EXPERT WORKSHOP ON CONGESTION MANAGEMENT
held on 29th-30th March 1993
at the Jefatura Provincial de Tráfico of Barcelona

Chairman: Mr. S. Strickland, United States
Host/Organiser: Mr. C. Lozano, Dirección General de Tráfico, Spain
Mr. A. Riu, Jefatura Provincial de Tráfico de Barcelona, Spain

INTRODUCTION: **CONGESTION MANAGEMENT PHILOSOPHIES/APPROACHES**

F. Schepis: Congestion Management Measures on Italian Inter-urban Motorways
Autostrade SpA, Italy

Y. Durand-Raucher: Traffic Management and Road User Information Service Ile-de-France
Traffic Operations Dept., Ile de France region   region on Motorways Around Paris (SIRIUS)

V. Himanen: Congestion Management Principles and their Application in the Helsinki
Technical Research Centre, Finland   Area

SESSION 1:   **CONGESTION MANAGEMENT IN AN URBAN/METROPOLITAN/REGIONAL**
SETTING: CASE STUDIES

T. Solheim: Introducing Urban Tolls : The Norwegian Experience
Institute of Transport Economics, Norway

A. Peterson: Experience with the LHOVRA Strategy for Traffic Signal Control of
Peek Traffic AB, Sweden   Isolated Intersections

U. Hammarström, VTI and Mr. A. Peterson: Experience with the AUT/TRANSYT System for Traffic Signal Control
Sweden   in a Network

S.O. Gunnarsson: Traffic Management Measures in Historic European Towns
Chalmers University of Technology, Sweden

M.G. Allan: Designation of a System of Priority (Red) Routes and Creation of the
Traffic Dir. for London, United Kingdom   Traffic Director for London

Mrs. T.E. de Bruijn: Integrated Regional & Traffic Planning Policy in Groningen
Rijkswaterstaat, Groningen Directorate
The Netherlands

137

J. Madsen:
City of Copenhagen, Denmark

The Copenhagen Experience: Minor Problems, Simple Solutions

N. Otten:
Daimler-Benz AG, Germany

Stuttgart Transport Operation by Regional Management (STORM):
Considerations for the Pilot Projects

A. Lindenbach:
Ministry of Transport, Hungary

Traffic Control Systems on the Hungarian Motorways

SESSION 2 : CONGESTION MANAGEMENT IN A CORRIDOR OR INTER-URBAN SETTING : CASE STUDIES

J.-Ch. Aquarone:
Swiss Federal Dept. of Transport

Combined Transport Objectives in Switzerland

F. Papon:
INRETS, France

Toll Modulation Experiment on the Lille-Paris Motorway (A1) on
Sundays since 1992: Implementation and Effects

R.H. de Boer:
Rijkswaterstaat, The Netherlands

Impact of a New Railway Line on Travel Patterns: The Flevo Line in
the Netherlands

J. Díez de Ulzurrun,
L. Serrano and A. Muñoz:
Dirección General de Tráfico, Spain

The Experience from the Seville, Barcelona and Madrid Traffic
Management Projects

D. Serwill:
Techn. University of Aachen, Germany

Holiday Traffic Forecasting in Germany

J. Slager:
Rijkswaterstaat, Utrecht Directorate
The Netherlands

The Amsterdam-Utrecht Corridor Study: Optimisation of the
Transport Infrastructure Supply

W. Berman:
Federal Highway Administration, United States

The Role of the National Incident Management Coalition in the U.S.

R. Lindenbach:
Road Management and Co-ordinating Directorate
Hungary

Traffic Control Systems on the Hungarian Motorways

SESSION 3: INSTITUTIONAL ISSUES IN FUTURE CONGESTION MANAGEMENT

A.I.J.M. van der Hoorn:
Rijkswaterstaat, The Netherlands

The Dutch 1986-2010 Transport Structure Plan: A Progress Note

C. Caubet:
SETRA, France

Institutional issues in French approaches

H. Dittmar:
Metropolitan Transportation Commission,
San Francisco Bay Area, United States

Implementing Congestion Management Programs: Confronting and
Resolving Institutional Barriers

LIST OF GROUP MEMBERS

Chairmen: Mr. Sheldon G. STRICKLAND (1993/94)
Mr. Steven C. LOCKWOOD (1992)
Vice Chairman: Mr. Cees J. LOUISSE
Technical Secretary: Mr. Wayne BERMAN

BELGIUM	Mr. M.Y. PARMENTIER
DENMARK	Mr. S. LAURITZEN
FRANCE	Mr. A. CARN
	Mr. F. PAPON
GERMANY	Mr. D. SERWILL
JAPAN	Mr. K. MASUDA
	Mr. K. MORI
	Mr. T. SAITO
	Mr. N. SATO
NETHERLANDS	Mr. C. LOUISSE
SPAIN	Mr. F. FERNANDEZ ALONSO
	Mr. F. GARCIA MATA
SWEDEN	Mr. S.O. GUNNARSSON
SWITZERLAND	Mr. G. PETERSEN
UNITED KINGDOM	Mr. R.J. BALCOMBE
UNITED STATES	Mr. W. BERMAN
	Mr. S. LOCKWOOD
	Mr. S. STRICKLAND
CEC	Mr. P. O'NEILL
OECD	Mr. B. HORN
	Mr. A. BARBAS
	Ms. V. FEYPELL

Rapporteurs of the Chapters of the Report were Messrs. W. Berman, A. Carn, F. Papon and D. Serwill. The final report was co-ordinated by the OECD Secretariat.

MAIN SALES OUTLETS OF OECD PUBLICATIONS
PRINCIPAUX POINTS DE VENTE DES PUBLICATIONS DE L'OCDE

ARGENTINA – ARGENTINE
Carlos Hirsch S.R.L.
Galería Güemes, Florida 165, 4° Piso
1333 Buenos Aires   Tel. (1) 331.1787 y 331.2391
                    Telefax: (1) 331.1787

AUSTRALIA – AUSTRALIE
D.A. Information Services
648 Whitehorse Road, P.O.B 163
Mitcham, Victoria 3132   Tel. (03) 873.4411
                             Telefax: (03) 873.5679

AUSTRIA – AUTRICHE
Gerold & Co.
Graben 31
Wien I   Tel. (0222) 533.50.14

BELGIUM – BELGIQUE
Jean De Lannoy
Avenue du Roi 202
B-1060 Bruxelles   Tel. (02) 538.51.69/538.08.41
                    Telefax: (02) 538.08.41

CANADA
Renouf Publishing Company Ltd.
1294 Algoma Road
Ottawa, ON K1B 3W8   Tel. (613) 741.4333
                         Telefax: (613) 741.5439
Stores:
61 Sparks Street
Ottawa, ON K1P 5R1   Tel. (613) 238.8985
211 Yonge Street
Toronto, ON M5B 1M4   Tel. (416) 363.3171
                         Telefax: (416)363.59.63

Les Éditions La Liberté Inc.
3020 Chemin Sainte-Foy
Sainte-Foy, PQ G1X 3V6   Tel. (418) 658.3763
                         Telefax: (418) 658.3763

Federal Publications Inc.
165 University Avenue, Suite 701
Toronto, ON M5H 3B8   Tel. (416) 860.1611
                         Telefax: (416) 860.1608

Les Publications Fédérales
1185 Université
Montréal, QC H3B 3A7   Tel. (514) 954.1633
                         Telefax : (514) 954.1635

CHINA – CHINE
China National Publications Import
Export Corporation (CNPIEC)
16 Gongti E. Road, Chaoyang District
P.O. Box 88 or 50
Beijing 100704 PR   Tel. (01) 506.6688
                         Telefax: (01) 506.3101

DENMARK – DANEMARK
Munksgaard Book and Subscription Service
35, Nørre Søgade, P.O. Box 2148
DK-1016 København K   Tel. (33) 12.85.70
                         Telefax: (33) 12.93.87

FINLAND – FINLANDE
Akateeminen Kirjakauppa
Keskuskatu 1, P.O. Box 128
00100 Helsinki
Subscription Services/Agence d'abonnements :
P.O. Box 23
00371 Helsinki   Tel. (358 0) 12141
                         Telefax: (358 0) 121.4450

FRANCE
OECD/OCDE
Mail Orders/Commandes par correspondance:
2, rue André-Pascal
75775 Paris Cedex 16   Tel. (33-1) 45.24.82.00
                          Telefax: (33-1) 49.10.42.76
                          Telex: 640048 OCDE
Orders via Minitel, France only/
Commandes par Minitel, France exclusivement :
36 15 OCDE
OECD Bookshop/Librairie de l'OCDE :
33, rue Octave-Feuillet
75016 Paris   Tel. (33-1) 45.24.81.67
                                  (33-1) 45.24.81.81
Documentation Française
29, quai Voltaire
75007 Paris   Tel. 40.15.70.00
Gibert Jeune (Droit-Économie)
6, place Saint-Michel
75006 Paris   Tel. 43.25.91.19
Librairie du Commerce International
10, avenue d'Iéna
75016 Paris   Tel. 40.73.34.60
Librairie Dunod
Université Paris-Dauphine
Place du Maréchal de Lattre de Tassigny
75016 Paris   Tel. (1) 44.05.40.13
Librairie Lavoisier
11, rue Lavoisier
75008 Paris   Tel. 42.65.39.95
Librairie L.G.D.J. - Montchrestien
20, rue Soufflot
75005 Paris   Tel. 46.33.89.85
Librairie des Sciences Politiques
30, rue Saint-Guillaume
75007 Paris   Tel. 45.48.36.02
P.U.F.
49, boulevard Saint-Michel
75005 Paris   Tel. 43.25.83.40
Librairie de l'Université
12a, rue Nazareth
13100 Aix-en-Provence   Tel. (16) 42.26.18.08
Documentation Française
165, rue Garibaldi
69003 Lyon   Tel. (16) 78.63.32.23
Librairie Decitre
29, place Bellecour
69002 Lyon   Tel. (16) 72.40.54.54

GERMANY – ALLEMAGNE
OECD Publications and Information Centre
August-Bebel-Allee 6
D-53175 Bonn   Tel. (0228) 959.120
                          Telefax: (0228) 959.12.17

GREECE – GRÈCE
Librairie Kauffmann
Mavrokordatou 9
106 78 Athens   Tel. (01) 32.55.321
                          Telefax: (01) 36.33.967

HONG-KONG
Swindon Book Co. Ltd.
13–15 Lock Road
Kowloon, Hong Kong   Tel. 366.80.31
                          Telefax: 739.49.75

HUNGARY – HONGRIE
Euro Info Service
Margitsziget, Európa Ház
1138 Budapest   Tel. (1) 111.62.16
                          Telefax : (1) 111.60.61

ICELAND – ISLANDE
Mál Mog Menning
Laugavegi 18, Pósthólf 392
121 Reykjavik   Tel. 162.35.23

INDIA – INDE
Oxford Book and Stationery Co.
Scindia House
New Delhi 110001   Tel.(11) 331.5896/5308
                          Telefax: (11) 332.5993
17 Park Street
Calcutta 700016   Tel. 240832

INDONESIA – INDONÉSIE
Pdii-Lipi
P.O. Box 269/JKSMG/88
Jakarta 12790   Tel. 583467
                          Telex: 62 875

ISRAEL
Praedicta
5 Shatner Street
P.O. Box 34030
Jerusalem 91430   Tel. (2) 52.84.90/1/2
                          Telefax: (2) 52.84.93
R.O.Y.
P.O. Box 13056
Tel Aviv 61130   Tél. (3) 49.61.08
                          Telefax (3) 544.60.39

ITALY – ITALIE
Libreria Commissionaria Sansoni
Via Duca di Calabria 1/1
50125 Firenze   Tel. (055) 64.54.15
                          Telefax: (055) 64.12.57
Via Bartolini 29
20155 Milano   Tel. (02) 36.50.83
Editrice e Libreria Herder
Piazza Montecitorio 120
00186 Roma   Tel. 679.46.28
                          Telefax: 678.47.51
Libreria Hoepli
Via Hoepli 5
20121 Milano   Tel. (02) 86.54.46
                          Telefax: (02) 805.28.86
Libreria Scientifica
Dott. Lucio de Biasio 'Aeiou'
Via Coronelli, 6
20146 Milano   Tel. (02) 48.95.45.52
                          Telefax: (02) 48.95.45.48

JAPAN – JAPON
OECD Publications and Information Centre
Landic Akasaka Building
2-3-4 Akasaka, Minato-ku
Tokyo 107   Tel. (81.3) 3586.2016
                          Telefax: (81.3) 3584.7929

KOREA – CORÉE
Kyobo Book Centre Co. Ltd.
P.O. Box 1658, Kwang Hwa Moon
Seoul   Tel. 730.78.91
                          Telefax: 735.00.30

MALAYSIA – MALAISIE
Co-operative Bookshop Ltd.
University of Malaya
P.O. Box 1127, Jalan Pantai Baru
59700 Kuala Lumpur
Malaysia   Tel. 756.5000/756.5425
                          Telefax: 757.3661

MEXICO – MEXIQUE
Revistas y Periodicos Internacionales S.A. de C.V.
Florencia 57 - 1004
Mexico, D.F. 06600   Tel. 207.81.00
                          Telefax : 208.39.79

NETHERLANDS – PAYS-BAS
SDU Uitgeverij Plantijnstraat
Externe Føndsen
Postbus 20014
2500 EA's-Gravenhage   Tel. (070) 37.89.880
Voor bestellingen: Telefax: (070) 34.75.778

NEW ZEALAND
NOUVELLE-ZÉLANDE
Legislation Services
P.O. Box 12418
Thorndon, Wellington   Tel. (04) 496.5652
                            Telefax: (04) 496.5698

NORWAY – NORVÈGE
Narvesen Info Center – NIC
Bertrand Narvesens vei 2
P.O. Box 6125 Etterstad
0602 Oslo 6   Tel. (022) 57.33.00
                          Telefax: (022) 68.19.01

PAKISTAN
Mirza Book Agency
65 Shahrah Quaid-E-Azam
Lahore 54000   Tel. (42) 353.601
                         Telefax: (42) 231.730

PHILIPPINE – PHILIPPINES
International Book Center
5th Floor, Filipinas Life Bldg.
Ayala Avenue
Metro Manila   Tel. 81.96.76
                         Telex 23312 RHP PH

PORTUGAL
Livraria Portugal
Rua do Carmo 70-74
Apart. 2681
1200 Lisboa   Tel.: (01) 347.49.82/5
                         Telefax: (01) 347.02.64

SINGAPORE – SINGAPOUR
Gower Asia Pacific Pte Ltd.
Golden Wheel Building
41, Kallang Pudding Road, No. 04-03
Singapore 1334   Tel. 741.5166
                         Telefax: 742.9356

SPAIN – ESPAGNE
Mundi-Prensa Libros S.A.
Castelló 37, Apartado 1223
Madrid 28001   Tel. (91) 431.33.99
                         Telefax: (91) 575.39.98

Libreria Internacional AEDOS
Consejo de Ciento 391
08009 – Barcelona   Tel. (93) 488.30.09
                         Telefax: (93) 487.76.59

Llibreria de la Generalitat
Palau Moja
Rambla dels Estudis, 118
08002 – Barcelona
        (Subscripcions) Tel. (93) 318.80.12
        (Publicacions) Tel. (93) 302.67.23
                       Telefax: (93) 412.18.54

SRI LANKA
Centre for Policy Research
c/o Colombo Agencies Ltd.
No. 300-304, Galle Road
Colombo 3   Tel. (1) 574240, 573551-2
                        Telefax: (1) 575394, 510711

SWEDEN – SUÈDE
Fritzes Information Center
Box 16356
Regeringsgatan 12
106 47 Stockholm   Tel. (08) 690.90.90
                         Telefax: (08) 20.50.21

Subscription Agency/Agence d'abonnements :
Wennergren-Williams Info AB
P.O. Box 1305
171 25 Solna   Tel. (08) 705.97.50
                         Téléfax : (08) 27.00.71

SWITZERLAND – SUISSE
Maditec S.A. (Books and Periodicals - Livres
et périodiques)
Chemin des Palettes 4
Case postale 266
1020 Renens   Tel. (021) 635.08.65
                         Telefax: (021) 635.07.80

Librairie Payot S.A.
4, place Pépinet
CP 3212
1002 Lausanne   Tel. (021) 341.33.48
                         Telefax: (021) 341.33.45

Librairie Unilivres
6, rue de Candolle
1205 Genève   Tel. (022) 320.26.23
                         Telefax: (022) 329.73.18

Subscription Agency/Agence d'abonnements :
Dynapresse Marketing S.A.
38 avenue Vibert
1227 Carouge   Tel.: (022) 308.07.89
                         Telefax : (022) 308.07.99

See also – Voir aussi :
OECD Publications and Information Centre
August-Bebel-Allee 6
D-53175 Bonn (Germany)   Tel. (0228) 959.120
                           Telefax: (0228) 959.12.17

TAIWAN – FORMOSE
Good Faith Worldwide Int'l. Co. Ltd.
9th Floor, No. 118, Sec. 2
Chung Hsiao E. Road
Taipei   Tel. (02) 391.7396/391.7397
                      Telefax: (02) 394.9176

THAILAND – THAÏLANDE
Suksit Siam Co. Ltd.
113, 115 Fuang Nakhon Rd.
Opp. Wat Rajbopith
Bangkok 10200   Tel. (662) 225.9531/2
                        Telefax: (662) 222.5188

TURKEY – TURQUIE
Kültür Yayinlari Is-Türk Ltd. Sti.
Atatürk Bulvari No. 191/Kat 13
Kavaklidere/Ankara   Tel. 428.11.40 Ext. 2458
Dolmabahce Cad. No. 29
Besiktas/Istanbul   Tel. 260.71.88
                           Telex: 43482B

UNITED KINGDOM – ROYAUME-UNI
HMSO
Gen. enquiries   Tel. (071) 873 0011
Postal orders only:
P.O. Box 276, London SW8 5DT
Personal Callers HMSO Bookshop
49 High Holborn, London WC1V 6HB
                         Telefax: (071) 873 8200
Branches at: Belfast, Birmingham, Bristol, Edin-
burgh, Manchester

UNITED STATES – ÉTATS-UNIS
OECD Publications and Information Centre
2001 L Street N.W., Suite 700
Washington, D.C. 20036-4910 Tel. (202) 785.6323
                            Telefax: (202) 785.0350

VENEZUELA
Libreria del Este
Avda F. Miranda 52, Aptdo. 60337
Edificio Galipán
Caracas 106   Tel. 951.1705/951.2307/951.1297
                        Telegram: Libreste Caracas

Subscription to OECD periodicals may also be
placed through main subscription agencies.

Les abonnements aux publications périodiques de
l'OCDE peuvent être souscrits auprès des
principales agences d'abonnement.

Orders and inquiries from countries where Distribù-
tors have not yet been appointed should be sent to:
OECD Publications Service, 2 rue André-Pascal,
75775 Paris Cedex 16, France.

Les commandes provenant de pays où l'OCDE n'a
pas encore désigné de distributeur peuvent être
adressées à : OCDE, Service des Publications,
2, rue André-Pascal, 75775 Paris Cedex 16, France.

11-1994

OECD PUBLICATIONS, 2 rue André-Pascal, 75775 PARIS CEDEX 16
PRINTED IN FRANCE
(77 95 02 1) ISBN 92-64-14315-7 - No. 47637 1994

Date Due

Mck DUE JAN 2 6 2004		
MCK RTD JAN 2 0 2004	MCK DUE SEP 3 0 2005	
CLIS RET SEP 2 9 2005		
OCT 3 1 2006		
MCK RTD OCT 3 1 2006		